SOMERSET UNDERGROUND

VOLUME 2
West Mendip Burrington and North Mendip

R. M. Taviner

Published by

MCRA

Mendip Cave Registry & Archive

2020 © R. M. Taviner

ISBN for the complete set of 4 volumes: 978-1-913271-00-8
ISBN for this volume: 978-1-913271-02-2

All rights reserved. No part of this book may be reproduced or transmitted in any form or by any means, electronic, graphic or mechanical, including photocopying, recording, taping, or by any other storage retrieval system (other than for purposes of review) without the express permission of the author or publisher in writing.

The right of Robin Mark Taviner to be identified as the author of this work, has been asserted by him in accordance with the Copyright, Designs and Patents Act 1988.

This book is sold subject to the condition that it shall not, by way of trade or otherwise, be lent, resold, hired out or otherwise circulated without the publisher's prior consent in any form of binding or cover other than that in which it is published and without a similar condition including this condition being imposed on the subsequent purchaser.

No responsibility is assumed by the publisher or the author for any injury and/or damage to persons or property as a matter of products liability, negligence or otherwise, or from any use or operation of any methods, products, instructions, information, or ideas contained in the material herein.

British Library Cataloguing in Publication Data
A catalogue record for this book is available from the British Library.

The book covers, and page layout designed by
Mark Gonzo Lumley

The Creative Edge

www.creativepembroke.co.uk

Printed and bound in India by Replika Press

Cover picture: Aveline's Hole by Mark Burkey
Rear cover picture: The Drainpipe, Goatchurch Cavern by Matt Voysey

Whilst every care has been taken in the compilation of this guide, neither the author nor the publisher can accept responsibility for any errors or omissions.

This volume is dedicated to the many
former students of Sidcot School
whose determination to live up to
their school motto did much to unlock
the inner secrets of Mendip's western hills

Live Adventurously!

Photos courtesy of Sidcot School and Wells and Mendip Museum

CONTENTS

- 8 **Introduction**
- 10 **History**
- 12 **Geology and Geomorphology**
- 16 **Archaeology and Palaeontology**
- 20 **Access**
- 22 **Water - A Precious Commodity**
- 24 **Complete or not Complete?**
- 25 **Descriptions**

WEST MENDIP

- 28 Brean Down
- 36 Uphill
- 42 Oldmixon
- 43 Hutton
- 60 Bleadon
- 63 Loxton
- 69 Banwell
- 82 Sandford
- 108 Churchill
- 111 Winscombe
- 114 Star & Shipham
- 126 Compton Bishop & Cross
- 136 Axbridge
- 152 Shipham Gorge

BURRINGTON

- 158 Dolebury & Rowberrow
- 163 Langford
- 165 Burrington Combe
- 195 Rickford

NORTH MENDIP

- 200 Blagdon
- 202 Ubley & Compton Martin
- 206 East Harptree
- 223 Eaker Hill & Greendown
- 234 Chewton Mendip

- 240 **Abbreviations**

   Appendices:
- 241 Appendix 1 - Mendip Cave Registry and Archive (MCRA)
- 242 Appendix 2 - Cave Conservation
- 244 Appendix 3 - Safety
- 246 Appendix 4 - Mendip Cave Rescue (MCR)
- 246 Appendix 5 - Caving Organisations
- 248 Appendix 6 - Further Reading
- 250 Acknowledgements
- 251 **Index**

INTRODUCTION

Somerset covers an area of 4,171 square kilometres and is a predominantly rural county of gently rolling hills separated by flat, fertile tracts of farmland and internationally important wetlands. There are no less than five major hill ranges, and these vary considerably, shaped by both the underlying geology, and the mild, but often damp and windy weather, which sweeps relentlessly in off the Atlantic Ocean. Unlike most of Britain, many of these upland areas were never subject to direct glaciation and therefore remain remarkably unaltered. The hills in the northern part of the county are composed chiefly of Carboniferous limestone, and karst features dominate - especially in the Mendip Hills, which is renowned for its caves and spectacular gorges. It is by some distance southern England's most important expanse of limestone. Immediately to the north lies the now defunct Somerset coalfield, and there is even a small outcrop of Silurian rock, which represents the oldest rock formation in the county. South of the Mendip Hills, the Polden Hills are composed of Triassic and Jurassic rocks, while south again the Blackdown Hills form the only extensive outcrop of Cretaceous Upper Greensand in the entire south-west of England. Turning west, towards the coast, the sandstones of the Quantocks mainly date from the Devonian period, and this theme continues into the Brendon Hills and on into Exmoor, which is home to both the tallest sea cliffs in mainland Britain, and Dunkery Beacon (520m A.O.D.), the county's highest point. Returning northwards, the coastline quickly drops to near sea level, before gradually rising up again as a series of low cliffs and narrow headlands which project out into the Bristol Channel. The result of all this complexity is a varied landscape of great beauty and interest, a status best reflected in the many levels of protection awarded to the county. These include one National Park (Exmoor), three Areas of Outstanding Natural Beauty (the Mendip Hills, the Quantocks, and the Blackdown Hills), fifteen National Nature Reserves, more than two hundred Sites of Scientific Interest, and countless other smaller reserves managed by local trusts.

Although this guide is laced with many fascinating snippets of local interest, it is the world below that remains the major focus of attention. Nature, of course, has been carving out natural caves and watercourses for millennia, but Somerset, historically, has also been an important centre for mining activity, and over the course of centuries almost any mineral you care to think of has been hauled out of the ground beneath our feet. Today, mining for minerals may have all but ceased, but the industry continues to thrive, now in the form of quarrying, an activity which although inherently destructive, frequently exposes features of scientific interest and importance that would otherwise have remained buried.

Cavers exploring the Somerset hills have long been served by excellent guide books. From *The Netherworld of Mendip*, (E.A. Baker, H.E. Balch), published in 1907, right through to the latest edition of *Mendip Underground*, (A. Gray, R. Taviner, R. Witcombe), published in 2013, a plethora of material has been made available, including the eponymous '*Mendip - The Complete Caves*' (N. Barrington & W. Stanton), fondly known as the 'Diggers' Bible'. Last published in 1977, this illustrious volume faithfully recorded details of more than 600 sites of speleological interest on or beneath the main Mendip plateau, and it's safe to say that its subtle hints of greater things to come inspired several generations of cave explorers on to new discoveries, including many of today's longest and most popular caves. Although superseded as a guidebook by Mendip Underground, and as a gazetteer by the online Mendip Cave Registry (**www.mcra.org.uk**), in many ways Complete Caves has never truly been replaced, and this guide - the second of four volumes covering the whole of Somerset - aims to address that gap. Many of the small sites mentioned here hold good potential, and it is hoped that like its illustrious predecessor, *Somerset Underground* will inspire both present and future generations of cave explorers on to yet more new discoveries.

Volume 1
Bristol, Broadfield Down, the Bristol Channel and West Somerset

Volume 2
West Mendip, Burrington and North Mendip

Volume 3
Central and South Mendip

Volume 4
East Mendip, Bath and South Somerset

HISTORY

Records pertaining to the caves and karst of Mendip date back to the 12th century, when Henry of Huntingdon numbered '*Cheddar Hole*' among his four great natural wonders of England. Then, in about 1676, lead miners broke into Lamb Leer Cavern, with its huge chamber, and the exploration of what was then the deepest recorded cave in the world generated considerable interest. Lead mining continued across the region into the 19th century, as did calamine extraction, a vital ingredient in the making of brass before an alternative and cheaper method of production was discovered. Ochre mining too proved of great importance, and this particular industry continued more or less continuously until well into the second half of the 20th century. These activities left considerable scars across much of the landscape covered by this guide, and a good deal remains visible today, notably around Banwell, Sandford Hill, Shipham, and on the hills above Axbridge. During the course of their endeavours the miners discovered several other notable caves, some of which contained important bone deposits. This attracted the attentions of antiquarian researchers, who gathered up large collections of ancient bones in their attempts to establish a connection with the Biblical flood. However, following the emergence of Charles Darwin's theories of evolution, a new

Coral Cave. Jack Duck collection, courtesy of MCRA

and more rigorous scientific approach to cave studies gradually began to emerge, epitomised by William Boyd Dawkins, who excavated several caves in Burrington Combe in the 1860s, publishing the results in his influential book *Cave Hunting* in 1874. This period also saw the rise of a wider public interest in the natural world, and inspired by this increase in activity, and by the lectures of William Boyd Dawkins, the 1890s saw H. E. Balch, and other members of the Wells Natural History and Archaeological Society, begin a systematic examination and exploration of the Mendip plateau. Mendip's first recognizable caving club, the Mendip Nature Research Committee (originally Mendip Nature Research Club), was founded in 1906, and the following year Balch and Ernest Baker, published *The Netherworld of Mendip*, which was the first English book solely devoted to caving. Soon, other cavers began to take an interest in the area, including the students at Bristol University, who would go on to form the University of Bristol Spelaeological Society in 1919. In 1925, schoolboy cavers from the Quaker Sidcot School near Winscombe began to enjoy success, and many of these youngsters went on to have distinguished caving careers with other clubs. This much needed influx of new blood prompted the establishment of clubs which placed an emphasis on the sporting aspects of caving, and the arrival of the Wessex Cave Club in 1934, and the Bristol Exploration Club the following year, led to a veritable explosion in Mendip caving. The first cave dives in Britain took place around this period, resulting in the exploration of several hitherto unknown passages and chambers. Other new clubs joined the fray, including the Shepton Mallet Caving Club in 1949, the Axbridge Caving Group in 1950, the Mendip Caving Group in 1954, and the Cerberus Spelaeological Society in 1956. Inevitably some of this attention turned towards West Mendip - especially Burrington Combe, a valley already renowned for the Mesolithic remains discovered in Aveline's Hole, and for Goatchurch Cavern, a well-known cave recorded as far back as 1736. Many more Burrington caves were entered in the years immediately following The Great War, and discoveries have occurred on a regular basis since, culminating in the finding of a series of well-decorated passages in Tween Twins Hole as recently as 2016. Further west, researchers have spent long hours attempting to rediscover several famous caves recorded by the miners and antiquarians, which were subsequently lost. Bleadon Cavern was finally re-entered in 1970, while the extensive calamine workings of Singing River Mine were regained shortly after. Shute Shelve Cavern was entered in the 1990s, followed by Loxton Cavern in 2003 (a deserved reward for some particularly clever detective work) and Axbridge Hill Cavern, with its chamber '*as big as Axbridge Square*', which was rediscovered in 2011. Extensive field work between 2014 and 2016 has shed considerable new light on several other lost caves on Sandford Hill (including the legendary Sandford Gulf), while the long-standing mystery surrounding perhaps the most famous lost cavern of them all (Hutton Cavern), was also finally solved in 2015, following intensive investigations around Canada Combe.

GEOLOGY AND GEOMORPHOLOGY

The north of Somerset is dominated by the Mendip Hills, a narrow tableland of mainly sedimentary rocks, which date from the Late Devonian (385-359 million years ago), right through the Carboniferous, Permian and Triassic to the Mid-Jurassic (161 million years ago). Stretching from Frome to the Bristol Channel, the spine of the hills comprises four large Devonian Old Red Sandstone periclines (overfolds), which were once more than 1000m in height, but now barely protrude above the general elevation. From east to west, these 'peaks' are the long, wooded whaleback ridge of Beacon Hill, Pen Hill with its giant television mast, North Hill, famous for its barrow cemetery, and heather-clad Blackdown, which at 325m forms the highest point on the hills. Surrounding these domes are a sequence of Carboniferous rocks, comprising Lower Limestone Shales overlain by several hundred metres of almost pure mountain limestone, together with localised outcrops of associated Millstone Grit and Coal Measures. In some places the overfolds were so intense that severe faulting occurred, further aggravating an already remarkably complex geological landscape. During the Triassic, screes comprising angular pieces of Devonian and Carboniferous material, accumulated in the ancient valleys, where they became loosely cemented with sand, producing Dolomitic Conglomerate, outcrops of which occur all over Mendip. To top it off, the entire range was once protected by a covering of Jurassic rocks, most of which has since been stripped away. This 'stripping' proceeded from west to east, and its early removal in the west has reduced the once-uniform plateau in this area to a series of fragmented rounded limestone hills, which descend gradually into the Bristol Channel. A corresponding lowering of the local water table left the caves exposed in the underlying limestones high and dry, which probably helps explain the preponderance of Pleistocene bone caves in the region - the bones often being found preserved within a matrix of windblown sands and ochreous material deposited during the Triassic.

When acidic rainwater falls onto limestone, a chemical reaction occurs which over time results in the removal of vast quantities of rock. Estimates for Mendip suggest that this solutional weathering accounts for the loss of roughly 40 cubic metres of limestone for every square kilometre of land area and the result is a mature karst environment, featuring typical landforms such as gorges, dry valleys and closed depressions, and a well-established pattern of predominantly subterranean drainage. To the north, the Mendip plateau terminates in a steep escarpment, punctuated by great gorges and steep-sided wooded combes, of which Harptree Combe,

Compton Combe, Burges's Combe and especially Burrington Combe are excellent examples. It was once thought that some of these gorges, were the result of cavern collapse, but it is now considered more likely that they formed rapidly during cold periods when water which normally flowed underground was forced back onto the surface. The relative sizes of these gorges, appears to be directly proportional to the size of their respective catchment areas. Cave collapse was once also thought to be responsible for Mendip's many closed depressions (often called dolines, sinkholes or shakeholes). Today however, they have been shown to be simply the result of the ground collapsing into localised solution joints, some of which have developed into small, localised cave systems. This augers well for future generations of cave explorers, for so far, of the many hundreds that are liberally spread across Mendip's surface plateau, only a mere handful have been seriously excavated.

Streams rising on the Old Red Sandstone run over the impervious Lower Limestone Shales to vanish on contact with the Carboniferous limestone in sink holes, known locally as swallets or slockers. The waters of these active cave systems combine to rise at a series of large springs located around the foot of the Mendips, of which the risings at Banwell, Langford, Rickford and Sherborne are prime examples. A number of these springs are tufaceous, and deposits have been dated to the Late Mesolithic (8000 BC - 4000 BC). Until relatively recently, not much was known about the links between Mendip's sinks and risings, but research conducted by caving organisations and professional bodies such as BARA and WWW, using dyed spores of Lycopdium clavatum (club moss), have established many positive traces, important information that has contributed greatly to our understanding of local cave development and the flow of water through Mendip's vast subterranean drainage network.

Ice Ages are periods that comprise alternating glacials and interglacials with associated stadials and interstadials. Stadials are relatively cold periods within an interglacial and, interstadials are relatively warm periods during glacials. They are classified using marine isotope stages (MIS). Derived from pollen and plankton remains found in deep sea cores, the levels of oxygen-18 indicate alternating warm and cool periods in the Earth's history. Caves can be a particularly valuable resource in this regard, offering a well-preserved environment in which these stages can be compared with uranium-series dates for speleothems to help provide information concerning both the cave's history and the region's paleoclimate. More than 100 stages have been identified, going back over millions of years, although comparisons are currently limited to around 500,000 years (the upper limit for uranium-series dating). The table below presents the delineations of Pleistocene glacials and interglacials. Odd numbers are warm periods and even numbers are cold periods.

GEOLOGICAL CHRONOLOGY
(Millions of years)

CENOZOIC
66 ma - present

Quaternary 2.6 ma – present
 Holocene 11,700 - present
 Pleistocene 2.6 ma - 11,700
Neogene 23 - 2.6 ma
 Pliocene 5.3 - 2.6 ma
 Miocene 23 - 5.3 ma
Palaeogene 66 -23 ma
 Oligocene 33.9 - 23 ma
 Eocene 56 - 33.9 ma
 Paleocene 66 - 56 ma

MESOZOIC
252 - 66 ma

Cretaceous 145 - 66 ma
Jurassic 201 - 145 ma
Triassic 252 - 201 ma

PALEOZOIC
541 - 252 ma

Permian 299 - 252 ma
Carboniferous 359 - 299 ma
Devonian 419 - 359 ma
Silurian 444 - 419 ma
Ordovician 485 - 444 ma
Cambrian 541 - 485 ma

PROTEROZOIC
2,500 - 541 ma

ARCHAEAN
4,000 - 2,500 ma

HADEAN
4,600 - 4,000 ma

BRITISH TERRESTRIAL STAGES

	Marine Isotope Stages (MIS)	

HOLOCENE

	warm	1

UPPER PLEISTOCENE
120 - 11.7 ka

Devensian	cold	2-5d
Ipswichian	warm	5e

MIDDLE PLEISTOCENE
781 - 120 ka

Wolstonian	cold	6-10
Hoxnian	warm	11
Anglian	cold	12
Cromerian	warm	13-20

LOWER PLEISTOCENE
2.588 ma - 781 ka

Beestonian	cold	21-62
Pastonian	warm	63-67
Baventian	cold	70-68
Antian/Bramertonian	warm	71-73
Thurnian	cold	74-76
Ludhamian	warm	75-80
Pre-Ludhamian	cold	81-100

ARCHAEOLOGY AND PALAEONTOLOGY

Archaeology is the study of peoples, past and present, through their material remains, whereas palaeontology, in the broadest sense of the term, refers to the study of fossils to determine the structure and evolution of extinct animals and plants and the age and conditions of the deposition of the rock strata in which they are found. The Pleistocene resources found within Somerset's caves and rock shelters are of international importance and continue to play a major role in the story of the ancient human occupation of Britain and our understanding of how people in the past lived. In the 18th and 19th centuries, antiquarians became increasingly interested in caves as a source of such material, and men like Alexander Catcott and William Beard, working on western Mendip, gathered large collections of ancient bones which led to much speculation about the effects of the Biblical flood. These collections contain rich assemblages of mammals including wolf, lion, cave bear, hyaena, bison, mammoth and woolly rhinoceros and many such bones can still be seen in situ in Banwell Bone Cave. Other finds can be seen in the Wells and Mendip Museum on Cathedral Green in Wells (with a particularly fine display in the Balch Room), while the very important William Beard Collection - the faunal remains of numerous Quaternary (Pleistocene) mammals, all gathered from the bone caves of western Mendip - now resides in the spectacular new Museum of Somerset in Taunton Castle. Smaller collections are to be found in the Wookey Hole Caves and Cheddar Caves museums, the North Somerset Museum in Weston-super-Mare, the Bristol City Museum, the University of Bristol Spelaeological Society's museum, as well as local history museums such as those in Axbridge, Bridgwater and Frome. A great deal of archaeological and palaeontological work was carried out by the early pioneers, especially H. E. Balch and E. K. Tratman, and today this work is carried on primarily by the UBSS, whose suitably-qualified experts regularly publish their findings in the society's Proceedings. Stratified deposits within caves offer a vital insight into the distant past and an alert eye is needed to spot archaeological finds before destruction by pick or spade. Study of such remains can reveal both details of past occupancy and important information on environmental conditions prevalent at the time. Cavers coming across bones or artefacts in the course of cave digging should seek expert help at an early stage.

PALAEONTOLOGICAL CHRONOLOGY

PLEISTOCENE

2.588 ma – 11.7 ka

Sites of Interest

Axbridge Ochre Middle Mine
Banwell Bone Cave
Banwell Ochre Caves
Blagdon Bone Fissure
Bleadon Bone Cave
Bleadon Cavern (Hutton Cavern)
Churchill Cave
Foxes Hole
Goatchurch Cavern
Loxton Cave
Picken's Hole
Pierre's Pot
Reindeer Rift
Triple Hole (Elephant Cave & Sandford Bone Fissure)
Triple-H Cave
Uphill Quarry Caves
Upper Canada Cave (Hutton Cavern 2)
Wolf Den

ARCHAEOLOGICAL CHRONOLOGY (Adapted from English Heritage)

PALAEOLITHIC
500,000 BC - 10,000 BC (Old Stone Age)
- **Lower Palaeolithic** 500,000 BC - 150,000 BC
- **Middle Palaeolithic** 150,000 BC - 40,000 BC
- **Upper Palaeolithic** 40,000 BC - 10,000 BC

MESOLITHIC
10,000 BC - 4000 BC (Middle Stone Age)
- **Earlier Mesolithic** 10,000 BC - 7000 BC
- **Later Mesolithic** 7000 BC - 4000 BC

NEOLITHIC 4000 BC - 2200 BC (New Stone Age)
- **Early Neolithic** 4000 BC - 3300 BC
- **Middle Neolithic** 3300 BC - 2900 BC
- **Late Neolithic** 2900 BC - 2200 BC

BRONZE AGE 2600 BC - 700 BC
- **Early Bronze Age** 2600 BC - 1600 BC
  - Beaker Period 2500 BC - 1700 BC
- **Middle Bronze Age** 1600 BC - 1200 BC
- **Late Bronze Age** 1200 BC - 700 BC

IRON AGE 800 BC to AD 43
- **Early Iron Age** 800 BC - 300 BC
- **Middle Iron Age** 300 BC - 100 BC
- **Late Iron Age** 100 BC - AD 43

ROMAN BRITAIN AD 43 - AD 410
Romano-British

EARLY MEDIEVAL AD 410 - 1066
Migration, Anglo-Saxon & Early Christian

MEDIEVAL 1066 - c.1540 (Dissolution)

POST-MEDIEVAL c.1540 - present
- **Industrial** 1700 - present
- **Modern** 1901 - present

ARCHAEOLOGICAL CHRONOLOGY

Sites of Interest

PALAEOLITHIC
500,000 BC - 10,000 BC

Rowberrow Cavern (possible)

MESOLITHIC
10,000 BC - 4000 BC

Aveline's Hole
Hay Wood Rock Shelter
Rowberrow Cavern (possible)

NEOLITHIC
4000 BC - 2200 BC
(New Stone Age)

Bos Swallet
Hay Wood Rock Shelter
Picken's Hole
Rowberrow Cavern
Uphill Quarry Caves

BRONZE AGE
2600 BC - 700 BC
(incl. Beaker Period)

Bos Swallet
Rowberrow Cavern

IRON AGE - ROMANO-BRITISH
800 BC to AD 410

Hay Wood Rock Shelter
Read's Cavern
Rowberrow Cavern
Scragg's Hole
Uphill Quarry Caves
Whitcombe's Hole

For a full overview of sites containing important archaeological material please refer to the article *'An overview of the archaeology of Mendip caves and karst'*, by V. Simmonds, which is freely available in the articles section of the MCRA website. Additional online information is available in *'Gazetteer of Caves, Fissures and Rock Shelters in Britain containing Human Remains'*, by A. T. Chamberlain, available from *http://caveburial.ubss.org.uk*

ACCESS

Land ownership changes frequently and given the sheer number of sites encompassed by this guide, it has not been possible to provide precise access arrangements. However, it should be remembered that practically every site is owned by somebody, and proper respect should be shown to the relevant organisations and individuals as it is only through their goodwill that such sites remain open for us all to explore and enjoy. Anyone visiting is strongly urged to seek proper permission. If necessary, please contact CSCC for the latest information. It goes without saying that visitors should observe the Countryside Code and not leave litter, damage property, pollute water supplies or leave gates open. If near property, please keep the noise down, especially after dark. Above all, be courteous to landowners and local people at all times!

Access to some caves is controlled by the Council of Southern Caving Clubs. These caves are usually locked, but a key common to all of them may be obtained from the major Mendip clubs. Alternatively contact the CSCC Conservation and Access Officer via the website **www.cscc.org.uk**. A number of other sites lie in nature reserves controlled by the Somerset Wildlife Trust. SWT has informed the caving community that visits to these caves and other sites may be made without seeking formal permission provided the Countryside Code and the simple rules applicable to the reserve are observed. However, permission to commence a dig (surface or underground) or an archaeological excavation MUST be obtained in advance. Other large organisations, such as Avon Wildlife Trust and The National Trust (NT), also own or lease significant tracts of land in Somerset, and although many sites remain readily accessible, access in some instances may be either denied or subject to special arrangements.

Sites of Special Scientific Interest (SSSIs) enjoy limited statutory protection under the provisions of the Wildlife and Countryside Act (1981). Local planning authorities must notify Natural England of any planned development which may cause damage to a scheduled SSSI, and landowners and tenants are precluded from carrying out certain potentially damaging operations specified by Natural England. A significant number of the sites encompassed by this guide are either scheduled as SSSIs in their own right, or fall within the boundaries of larger area karst, botanical or faunal SSSIs. Definitive maps are available on the Natural England website **www.naturalengland.org.uk**.

SSSIs within the scope of this volume include:

Axbridge Hill and Fry's Hill

Banwell Caves (Banwell Bone and Stalactite Caves)

Banwell Ochre Caves

Bleadon Hill

Brean Down

Burrington Combe

Cheddar Wood

Compton Martin Ochre Mine

Crook Peak to Shute Shelve Hill

Dolebury Warren

Harptree Combe

Lamb Leer

Max Bog (Winscombe)

Purn Hill

Shiplate Slait

The Perch

Uphill Cliff

Wurt Pit and Devil's Punchbowl

Permission for any surface digging within the confines of a scheduled SSSI MUST be sought from Natural England. In the interests of cave conservation, anyone noticing any activity that may be detrimental to an SSSI (above or below ground) should bring it to the attention of the CSCC Conservation and Access Officer. Further sites fall within the boundaries of other protected jurisdictions, such as National Parks, Areas of Outstanding Natural Beauty and National Nature Reserves, and permission to excavate must again be sought from the relevant authorities.

Finally, despite claims that most vehicle crime is preventable, the sad truth is that cars parked in isolated spots are particularly vulnerable to break-ins. Thieves often know where cavers and mine enthusiasts are most likely to leave their cars, and that they will probably not be returning to their cars for at least a couple of hours, which unfortunately leaves plenty of time for them to perform their nefarious activities. Please don't leave anything valuable on display.

WATER - A PRECIOUS COMMODITY

It's hard to underestimate the fundamental importance of water. After all, each and every one of us is made up of roughly 60% of the stuff, without which we would very quickly shrivel up and expire. Our forebears understood this all too well and harnessed many of our natural water sources to improve the longevity and quality of their lives. This fact is reflected not only in the place names of many of Somerset's most important settlements, including the cities of Bath and Wells, but also in their physical locations, such as on Mendip where the central plateau is encircled by villages constructed along the natural spring line. A corresponding lack of water on the surface plateau also helps explain the paucity of settlements on the higher ground, and this clearly dates back into ancient times, as demonstrated by the scarcity of important buildings, such as large mansions or Roman villas. A healthy proportion of our springs are also rich in important minerals, and several were thought capable of curing all manner of interesting complaints and ailments. Given that doctors (including those at the Bristol Eye Hospital), were prescribing natural spring water as recently as the founding of the NHS, it seems hardly surprising that our ancient ancestors revered such places as sacred, a fact not lost on the Christian church, who subsequently enclosed many of the most popular springs in elaborate stone structures to prevent contamination of the holy 'baptismal' waters. There are many such sacred springs and holy wells in the area covered by this guide, many of which were faithfully recorded in two excellent guides produced by Dom. E. Horne in 1923, and Phill Quinn in 1999. Those regarded as the most relevant, or important, have been included, along with some additional information.

Somerset is one of those parts of the world where water falls from the sky with monotonous regularity. So monotonous in fact, that we all but take it for granted. Yet it was not always so, and it may not always be this way in the future. In fact, this vital resource is under constant threat. The lead miners', both at Charterhouse and Priddy, were once responsible for causing widespread pollution of our underground watercourses, and there are numerous documented cases of lead contamination causing the death of livestock and fish downstream of springs and risings. This contamination also posed problems for Mendip's many paper mills, whose requirement for pure clean water often dictated that they be located as close to the risings as possible, and it wasn't until 1863 that this issue was finally resolved, when the owners of Wookey Hole mill successfully gained an injunction against the St Cuthbert's Lead Works, preventing them from polluting streams adjacent to their surface works. Nitrates too have caused issues. While these occur naturally in both soil and water, excess levels can be a contaminant, and most sources can be traced back to human activity, such as agricultural activities, human waste disposal or industrial pollution. For example, sewage, or fertilisers applied to open ground, can be washed into streams and rivers, causing algae and other aquatic plants to grow at an accelerated rate. The result is extreme fluctuations in dissolved oxygen, and this can have devastating consequences for invertebrates, fish and shellfish, while in the very worst-case scenarios it can even interfere with the ability of human red blood cells to transport oxygen, leading to breathing difficulties, especially in infants. In limestone areas, these effects are often exacerbated, for runoff can easily find its way into swallets, which provides a fast and more direct route for contaminated water to reach the risings, many of which remain important water supplies. Limestone areas are also particularly vulnerable to the effects of quarrying, where dewatering can affect, or even prevent ground water from reaching long-standing water supplies. This process is visible today at the Seven Springs and Holwell catchments, where pumping and intensive extraction at nearby super quarries has caused serious side effects. Additional problems can occur when dewatering ceases, for rising water levels can pick up contaminants that are quickly transported to underground aquifers, a situation especially hazardous if quarries are later reused as landfill sites. Other potential sources of contamination can include leaking storage and septic tanks, toxic substances from old and active mining sites, car oil, diesel and even the salt deposited on the roads in winter. All have been detected underground, and in recent years another potential threat to our water supplies has reared its ugly head, in the form of fracking, a process whereby chemicals are injected into the ground at high-pressure in order to fracture the underlying shale rocks in an attempt to extract the natural gases trapped within. This process has been proven to contaminate water supplies in some parts of the United States, while several other areas were forced to order emergency shut-downs of injection sites out of fears that contaminated fracking fluids were reaching the drinking water aquifers. Several exploratory licenses have been issued across Somerset, including one which centred on an area so geologically complex that the early geological surveyors recorded *'an amount of confusion and distortion which literally baffles description.'* Unhappily, this same area is also home to several important springs, which together make up a sizable proportion of the region's water supply. Fed by both fast-flowing swallet water, and slower-moving percolation water, the water travels along subterranean paths that are simply impossible to predict with any real certainty. To nobody's great surprise, recent license transgressions appear to have temporarily curtailed this threat, but any attempt to resurrect this potentially destructive industry should not be pursued until the longer-term implications to our vital water resources are considerably better understood.

COMPLETE OR NOT COMPLETE?

That is indeed the question. Any sensible analysis will quickly conclude that complete can never truly be complete, and the founding fathers of the MCRA recognised as much when stating *'the work of compiling the Registers will never be complete whilst cave exploration continues and publications are produced'*. We were only too well-aware when we first published online, that many of the sites listed on the registry were inaccurately placed, while a good many others were missing altogether. The process of fixing these issues has taken almost a decade to come to fruition and involved not just revisiting every practicable site listed on the registry, but also walking as many inches of ground as possible in between. The result of all those labours is *Somerset Underground*, a four-volume, snapshot in time publication of (hopefully!) accurate and reliable information, designed specifically for use in the field. The exhaustive fieldwork that went into producing it has yielded several hundred previously unrecorded sites, at least some of which must have been entered before but never properly recorded. And the process doesn't stop there of course for the world beneath our feet continues to evolve, and there must still be many other sites of interest out there that have yet to come to the attention of either the MCRA, or the wider caving community. Mineshafts in particular continue to appear with monotonous regularity, and over the centuries, hundreds, probably thousands, of mines have been lost to us, either grubbed out to make way for forestry and agriculture or buried beneath modern housing estates. Many were capped haphazardly, often in a very short space of time, and these can collapse open at any time.

Notwithstanding the stated aim of the MCRA to document *'any natural or artificial hole, swallet, sink, rising, spring and mine (other than coal mines)'*, certain aspects clearly need to be selective. While springs and risings of particular speleological interest have naturally been included, it's obviously a near-impossible task to document every single water source in the area covered by these guides. There are simply far too many, the majority of which are frankly of little interest. However, there are some that over the centuries have been prized for their mystical or curative properties, especially those sites rich in minerals, such as the red iron-stained chalybeate springs which our ancestors often associated with blood and death. Many such sources were later adopted by Christians, who dedicated them to locally-important saints. These 'holy wells' operated right up until the reformation, at which point their religious association was repressed, and their use declined. Others were turned into provincial health spas, and those sites considered the most important have also been included.

The purpose of the MCRA is to provide an accurate and readily-available central reference point for all information pertaining to the sites listed in the registry. The information within this guide has been meticulously researched, with special attention paid to filtering out any inaccurate information which may have inadvertently appeared in earlier guides. However, there is always room for improvement so any reliable information regarding omissions, inaccuracies, or an earlier (or different) provenance than that provided, would be gratefully received.

DESCRIPTIONS

Unlike Mendip Underground, in which caves and mines are presented in alphabetical order, sites described in this volume are grouped together within geographical areas. Where known, each site is provided with the usual 'vital statistics' - 8-figure National Grid Reference, altitude, length and vertical range (rather than simple depth). In the case of 'lost' caves or grubbed out mines, some of the figures quoted are estimations.

Due to space limitations, references for each site are usually limited to just one or two entries - those the author considers the best, or sometimes the only one available. For a more scientific synopsis, readers are frequently directed towards Barrington & Stanton's 'Complete Caves', while in the case of the larger systems, considerably more detail can be gleaned from Mendip Underground. Readers however, are strongly urged to research the comprehensive list, available in the bibliography section on the MCRA website (**www.mcra.org.uk**).

Somerset Underground · Volume 2

The Mendip Hills are one of England's truly special places, a landscape of gentle, undulating limestone uplands, punctuated by enigmatic caves and rugged gorges. Ancient monuments abound and the views over the Somerset Levels to the south and the Chew Valley to the north have been ranked among the finest in England. In 1972, 198 square kilometres of this, the most southerly Carboniferous limestone upland in Britain, was designated as an Area of Outstanding Natural Beauty (AONB), a status equivalent to a National Park and intended to conserve and enhance the natural beauty of the environment and protect it from intrusive development. The spine of the Mendips is formed along an east-west axis, which according to the bounds of the ancient Saxon royal hunting forest, begins at a place called Kotellisasch (Cottle's Oak) on the outskirts of Frome. From here the hills rise gently towards the highest point on Blackdown (325m AOD), before descending

WEST MENDIP

No piece of country in the kingdom offers so much to explore. An abundant harvest is there waiting to be reaped; for on every side are obvious indications of half-buried gateways to the dark and secret pathways to the netherworld, and everywhere upon the surface of the Mendip tableland lie the open pits and hollows which the local speech calls swallets ... all testifying to untold ages of water action. **The Netherworld of Mendip**

towards the Bristol Channel in a series of low hills which terminate at 'le Blacston' (Black Rock), a tiny rocky islet near the mouth of the River Axe at the eastern end of Brean Down. Beneath these hills lie many hundreds of natural caves and in some areas the rock has become mineralised with lead and zinc ores, or infilled with ochreous deposits, making the hills a historically important centre for mining activity. The main focus of this particular volume concentrates on the range's western hills which are particularly rich in caves of both historical and archaeological importance. Loosely speaking, these small individual hills can be grouped into two distinct arms, flanking the Lox Yeo river. The northern arm comprises Hutton, Banwell, Sandford and Shipham, while the southern arm overlooks the villages of Bleadon, Loxton, Compton Bishop and Axbridge. The hills terminate in the Bristol Channel at the prominent limestone headland of Brean Down.

BREAN DOWN

The prominent limestone headland which separates Weston Bay from Bridgwater Bay, derives its name from a pairing of the Celtic (bryn), meaning hill, with the Saxon equivalent, an unusual etymological combination repeated further east at Crook Peak and Pen Hill. Standing almost 100m high, it extends 1.5km out to sea and is surrounded by steep cliffs. On the south side there is a remarkable sand cliff containing human burials and palaeontological evidence dating back to the last ice age. Home to both a small Romano-British temple and an impressive Victorian Palmerston fort, the headland was used by Marconi to set a new distance record for wireless transmissions over open sea. A garrison was stationed here during WWII and the curious small railway beneath the fort was used by Admiralty staff based on Birnbeck Island (HMS Birnbeck), to launch experimental rockets out across the Bristol Channel. Brean Down is also the suggested starting point for the Severn Barrage, a gigantic and ambitious hydro-electric scheme which thankfully has yet to come to fruition. The caves, the bulk of which were recorded by the MCRA during the making of this guide, are described in a clockwise direction beginning on the south coast of the headland. By timing the tides correctly (i.e. following the receding tide in this direction), it is possible to circumnavigate the headland and visit all the caves on the point in one session. However please be advised that all the caves flood completely at high tide and escape to the headland above (particularly from the south side) can be difficult if not impossible in places. The tide can come in very fast and the flats are highly dangerous to those who don't know how to cross them. Please take great care. Note - as the headland projects into the Bristol Channel, Brean Down, along with neighbouring Uphill, are also included in Volume One of this series (Bristol, Broadfield Down, the Bristol Channel and West Somerset).

REINDEER RIFT (ST 2945 5876) L 15m VR 15m

Also known as Cyclops Cave, Reindeer Fissure and Raven's Cave, this is a very large and impressive sea cave, which terminates in a large choke of jammed thermoclastic rubble from which reindeer antler and rodent bones have been extracted. These deposits probably originated through a choked blowhole in the roof. A second small hole (L 3m), located in a steep gully directly above the cave and close to the top of the cliff, can only be reached from above. A famous sand cliff, located 100m to the east, has yielded a wealth of archaeological material relating to a sequence of four main prehistoric occupation layers.

Ref: A. M. ApSimon, D. T. Donovan & H. Taylor, The Stratigraphy and Archaeology of the Late-Glacial and Post-Glacial Deposits at Brean Down, Somerset, UBSS Proc 9 (2) pp 88-90 (1960-61)

MCRA-BD-1 (ST 2913 5875) L 10m VR 5m

This sea cave contains a curious white rock formation at the rear and a narrow blowhole and aven on the left.

Great Rift. Photo *by* Rob Taviner

MCRA-BD-2 (ST 2911 5876)   L 8m VR 4m
A narrow rift with a very tight slot to daylight on the left and an impassable blowhole.

MCRA-BD-3 (ST 2910 5876)   L 2m VR 1m
4m west of MCRA-BD-2 is a very narrow rift.

MCRA-BD-4 (ST 2907 5877)   L 5m VR 5m
An almost entirely unroofed sea cave.

MCRA-BD-5 (ST 2877 5880)   L 6m VR 3m
A roomy arch leads to a standing-sized rift.

GREAT RIFT (ST 2823 5900)   L 20m VR 7m
Also logged as MCRA-BD-6, this huge cleft is very similar in appearance to Reindeer Rift. A second large entrance can be reached by a sloping ledge and there is a stratified choke fully 20m high, in which what appear to be bones are clearly visible.

MCRA-BD-7 (ST 2811 5906)   L 4m VR 2m
Situated 3m above the foreshore at the rear of a fine wave-cut platform, this is a narrow rift formed inside a mineral vein.

MCRA-BD-8 (ST 2810 5908)   L 3m VR 2m
A short scramble to the right of Giant's Cave yields a short rift.

GIANT'S CAVE (ST 2809 5909)   L 10m VR 5m
Also logged as MCRA-BD-9, this is the largest cave entrance on the headland (5m x 5m), and can be approached either by a climb over, or narrow passage beneath, huge wedged boulders. Sadly, the cave swiftly degenerates to a choke.

MCRA-BD-10 (ST 2803 5913)   L 2m VR 1m
Little more than an alcove.

LOOKOUT CAVE (ST 2800 5918)   L 10m VR 5m
Also logged as MCRA-BD-11, this sea cave is situated directly below the observation post to the south side of the fort. A high narrow rift ascends a series of short steps to reach a climb up into an upper chamber containing a blowhole.

DITCH CAVE (ST 2800 5920)   L 12m VR 5m
Also logged as MCRA-BD-12, this large and attractively-sculpted sea cave is situated directly below the ditch retaining wall protecting the land side of the fort.

POOL CAVES (ST 2798 5926)   L 2m VR 1m
Also logged as MCRA-BD-13, these are two short caves situated by a rock pool on a rocky ledge 10m above shore level.

Steep Holm from Ditch Cave. Photo by Rob Taviner

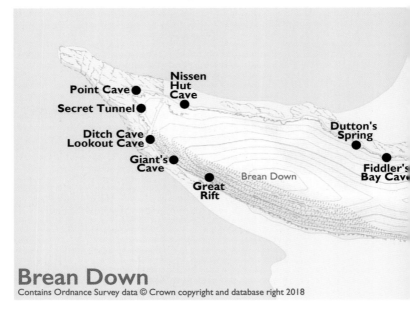

Brean Down

Contains Ordnance Survey data © Crown copyright and database right 2018

MOAT CAVE (ST 2795 5929) L 8m VR 2m
Also logged as MCRA-BD-14, this is a beautifully-sculpted unroofed sea cave with notable solutional features.

SECRET TUNNEL (ST 2800 5928) L 10m VR 2m
Also logged as MCRA-BD-15, this unusual cave is situated high and dry at the rear of a long inlet well above shore level. It appears to be partially-mined and contains an old drain, piped into it from the fort above.

POINT CAVE (ST 2797 5936) L 4m VR 2m
Also logged as MCRA-BD-16, this is the most westerly sea cave on the headland and is an attractively-sculpted passage with a blowhole.

MCRA-BD-17 (ST 2804 5935) L 2m VR 2m
Little more than an overhang below the north-west corner of the fort.

NISSEN HUT CAVE (ST 2815 5929) L 7m VR 4m
Also logged as MCRA-BD-18, a crawl beneath a massive concrete block yields a narrow rift with a connecting hole through to the foundations of a WWII Nissen hut.

MCRA-BD-19 (ST 2826 5928) L 2m VR 2m
A tiny cave with a blowhole.

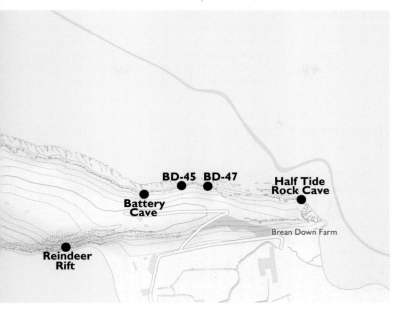

MCRA-BD-20 (ST 2833 5926) L 4m VR 2m
The only one of several adjacent inlets that is 'roofed over' this is an attractive twisting pebble-floored crawl.

MCRA-BD-21 (ST 2851 5922) L 3m VR 2m
A narrow inlet leads to a pleasant but equally narrow rift. There are four unusual holes in the cliff face directly above.

MCRA-BD-22 (ST 2865 5922) L 4m VR 1m
A narrow crawl at the west end of Fiddler's Bay.

MCRA-BD-23 (ST 2866 5920) L 2m VR 1m
The mere beginnings of a sea cave!

MCRA-BD-24 (ST 2869 5918) L 5m VR 3m
A roomy but short through cave.

MCRA-BD-25 (ST 2876 5914) L 10m VR 8m
A high roomy rift.

MCRA-BD-26 (ST 2877 5913) L 3m VR 1m
A large inlet leads to a low crawl through a pool.

DUTTON'S SPRING (ST 2877 5913)

Named after Louis M. Dutton, who recorded it in his pamphlet on Brean Down c.1921, this spring (the only natural source of fresh water on the headland) is otherwise known as Brean Down Resurgence. Located roughly 50m west of Fiddler's Bay Cave, it dries up completely in summer. A cap of red brickwork constructed to divert the water into a small rock basin can no longer be identified.

Ref: L. M. Dutton, Brean Down (1921)

FIDDLER'S BAY CAVE (ST 2882 5910)   L 16m VR 6m

This roomy cave appears to be largely phreatic in origin and contains a number of notable solution hollows. Lumps of yellow ochre can be seen jammed into cracks and crevices in the floor.

Ref: N. Richards & N. Harding, The Caves on Brean Down, BEC Belfry Bull 523 pp 22-27 (2005)

MCRA-BD-27 (ST 2888 5908)   L 2m VR 1m
A very small hole at the rear of a disproportionately large inlet!

MCRA-BD-28 (ST 2890 5908)   L 2m VR 1m
Another very short and narrow hole.

MCRA-BD-29 (ST 2900 5909)   L 3m VR 2m
A clamber over seaweed encrusted boulders to the right of a large inlet yields a short rift.

MCRA-BD-30 (ST 2903 5910)   L 4m VR 1m
A flat-out crawl over cobbles soon becomes too tight to pursue. It is recognisable by a tyre jammed inside (2015).

MCRA-BD-31 (ST 2905 5910)   L 3m VR 1m
Another low crawl over pebbles.

MCRA-BD-32 (ST 2906 5911)   L 7m VR 3m
A high attractive rift which terminates in a pile of boulders.

MCRA-BD-33 (ST 2907 5912)   L 2m VR 1m
The westernmost of three wave-cut igloo-shaped alcoves.

MCRA-BD-34 (ST 2909 5915)   L 7m VR 6m
A high narrow rift with a sculpted floor and blowholes.

MCRA-BD-35 (ST 2922 5912)   L 2m VR 1m
A low crawl over pebbles.

MCRA-BD-36 (ST 2925 5909)   L 5m VR 2m
A large inlet leads to a short through cave.

MCRA-BD-37 (ST 2932 5909)   L 8m VR 3m
A narrow attractive rift with a large blowhole.

MCRA-BD-38 (ST 2948 5902)   L 2m VR 1m
A very short narrow rift.

MCRA-BD-39 (ST 2955 5898)   L 3m VR 1m
A flat-out pebble crawl roofed entirely with jammed boulders.

MCRA-BD-40 (ST 2958 5897)   L 3m VR 1m
A short narrow sea cave.

BATTERY CAVE (West) (ST 2964 5894)   L 54m VR 15m
This impressive cave is located in a large embayment in the cliffs, a short distance west of a row of WWII gun emplacements. The western entrance, which lies on the western edge of the cove, is a very high, draughting rift which leads directly to an extensive and locally well-decorated natural bedding plane. A passage to the left leads to the alternative, eastern entrance.

Ref: N. Richards & N. Harding, The Caves on Brean Down, BEC Belfry Bull 523 pp 22-27 (2005)

BATTERY CAVE (East) (ST 2965 5893)
The alternative entrance to Battery Cave is located centre-right at the rear of the small cove.

MCRA-BD-41 (ST 2967 5896)   L 3m VR 2m
A roomy entrance which diminishes immediately.

MCRA-BD-42 (ST 2969 5896)   L 3m VR 2m
A roomy alcove with a small blowhole.

MCRA-BD-43 (ST 2970 5896)   L 3m VR 2m
Situated immediately east of MCRA-BD-42 this is another low crawl with a blowhole.

MCRA-BD-44 (ST 2971 5896)   L 2m VR 1m
A low cave which unusually runs parallel to the sea.

MCRA-BD-45 (ST 2979 5897)   L 20m VR 6m
A large entrance leads into a roomy chamber with a large high-level blowhole. A narrow rift at the rear can be pursued for 5m until it becomes too tight.

MCRA-BD-46 (ST 2980 5897)   L 5m VR 2m
A roomy sea cave.

MCRA-BD-47 (ST 2988 5896)   L 10m VR 4m
The entrance lies in a large inlet and is obscured by fallen boulders. A drop into a pool yields a roomy dark twisting rift which all too soon degenerates to a crawl.

MCRA-BD-48 (ST 2990 5896) L 8m VR 3m
A high narrow rift.

MCRA-BD-49 (ST 2991 5897) L 8m VR 2m
A beautifully-sculpted sea cave.

MCRA-BD-50 (ST 2994 5896) L 8m VR 2m
A narrow sea cave.

MCRA-BD-51 (ST 3001 5894) L 7m VR 2m
A narrow sea cave with pools.

MCRA-BD-52 (ST 3004 5894) L 3m VR 1m
A very small hole.

MCRA-BD-53 (ST 3007 5892) L 2m VR 1m
A low bedding at the rear of a large inlet.

MCRA-BD-54 (ST 3010 5894) L 5m VR 2m
A crawl with a blowhole.

MCRA-BD-55 (ST 3011 5894) L 3m VR 1m
A narrow rift.

MCRA-BD-56 (ST 3018 5893) L 2m VR 2m
A small wave-cut shelter.

HALF TIDE ROCK CAVE (ST 3021 5893) L 31m VR 6m
This large inclined bedding cave gradually diminishes in size before emerging from a second entrance located in the next cove to the west. There is some flowstone present.

Ref: N. Richards & N. Harding, The Caves on Brean Down, BEC Belfry Bull 523 pp 22-27 (2005)

UPHILL

The village of Uphill is situated on the River Axe at the south end of Weston Bay. Once an important Roman port (used for shipping the lead extracted from the Mendips), it is now the starting point for the Mendip Way, an 80km footpath that encompasses the entire Mendip range before terminating in Frome. To the south of the village there is a steep hill, topped by the remains of a 14th century windmill, and the unroofed church of St Nicholas, itself built on the site of a Romano-British temple. There is evidence of small-scale mining near the windmill (probably galena), and quarrying of this hill between the 18th and 20th centuries revealed several small caves, some of which produced important archaeological material. Most were either buried or destroyed, and today only three fragments remain, although other caves may yet await discovery beneath the parts of the hill unaffected by quarrying. The area is now protected as the Uphill Cliff SSSI. The caves are numbered in the order of discovery, with some of the NGRs inferred using material gleaned from old maps and reports.

UPHILL CAVERN (ST 3153 5841)   L 50m VR 20m Alt 16m

Also known as Uphill 1, Uphill Quarry Cave 1 or Williams' Cave, this, the first and most famous of the thirteen caves discovered in Uphill Quarry, was intercepted in 1826. The cave contained Pleistocene bones, including those of rhinoceros, bear, horse and hyaena, along with many smaller mammals. Most of the bones were gnawed, and the hyaena remains were so numerous that the fissure was almost certainly a hyaena's den. Under the guidance of Rev. David Williams, two workmen pursued a fissure downwards until it broke into a cavern described as 13m long, 7m wide and up to 4m high, and floored with an iron-stained sandy fill covered with a thin layer of mud. This was thought to have been deposited by a relatively recent inundation from the sea, possibly even the great flood of 30 January 1607, which swept away entire villages and killed thousands of people and animals. Roman pottery and two coins, one dating from the period of the emperor Julian (AD 133), were discovered in a side

fissure, along with sheep bones and material usually only attributed to foxes. Surprisingly, no further Pleistocene remains materialised, but E. C. H. Day, writing in 1866 attributed this to the fact that the lower portion of the cave may well have been filled with sea water during the relevant period. A second entrance (Uphill Quarry Cave 2) into this chamber was later opened up, while side passages in the upper part of the original entrance fissure yielded another chimney entrance and a further connection with the main chamber. Williams sent bones to the Bristol Institution (which later became the Bristol Museum and Art Gallery), the antiquarian Professor William Buckland, and to the Geological Society in London. Most are now lost but the British Museum retains several attributed to here. Stanton suggested that Uphill Quarry Cave 12 might be a surviving remnant, but this seems impossible for old maps clearly mark the location of the cave some 90m away at what E. Wilson and S. H. Reynolds (1898) described as the north end of the quarry. The bulk of Uphill Cavern was probably completely quarried away sometime between 1863 and 1898.

Ref: J. Rutter, Delineations of the North Western Division of the County of Somerset, Longman Rees & Co. pp 78-83 (1829) & R. A. Harrison, The Uphill Quarry Caves, Weston-super-Mare, A Reappraisal, UBSS Proc 14(3) pp 233-254 (1977)

UPHILL QUARRY CAVE 2 (ST 3152 5841) Alt 1m

Also recorded as Uphill 2, the entrance to this cave lay at quarry floor level near the lime kiln and was initially blocked with a bank of sand and loose stones. It was unblocked to allow easier access into the main chamber of Uphill Cavern, and a pot containing 129 silver and copper Roman coins dating from AD 364-375 was unearthed during the excavation. This lay beneath a large roof fissure, through which it had presumably fallen. The cave was later used as an explosives store, complete with blast doors, which were still in evidence in 1898, and Balch, writing in 1936, recorded that a small part of it still existed, albeit nearly buried. It was probably destroyed shortly after, but it's just possible that a fragment may still survive, buried beneath tipped material filling the quarry floor.

Ref: Cavern at Uphill, The Gentleman's Magazine, Vol. 26, 180 p 633 (1846)

UPHILL QUARRY CAVE 3 (ST 3151 5843) L 12m VR 5m Alt 21m

Also known as Pooley's Cave and Uphill 3, this site was discovered by James Parker in 1863. Several crania were apparently removed to a museum '… *in Oxford*', before it was excavated by Charles Pooley, who discovered further human remains, alongside a charcoal deposit, a sherd of pottery, and the bones of many animals, including wolf, wild cat, boar and deer. The human remains were probably Neolithic in origin, and quarrying had completely destroyed the cave by 1900.

Ref: C. Pooley, Notes and Queries, The Geologist 6 p 231 (1863)

UPHILL QUARRY CAVE 4 (ST 3156 5836)   Alt 12m
A short cave visited by James Parker who noted red breccia. It has been quarried away.

UPHILL QUARRY CAVE 5 (ST 3160 5829)   Alt 8m
Another small cave recorded by James Parker. No further details are known, and the cave has been quarried away.

UPHILL QUARRY CAVE 6 (ST 3159 5833)   Alt 10m
Also recorded as Uphill 6, this cave was discovered in 1881 *'high up in the face of the quarry'*. The cave had a broken stalagmite floor, and F. A. Knight reported finding fragments of a human skull and some charcoal underneath. Judging by the worn nature of the teeth, Knight considered the skull to be that of an old man. The cave has since been quarried away.

Ref: F.A. Knight, The Sea-Board of Mendip, J. M. Dent & Company (1902)

UPHILL QUARRY CAVE 7 (ST 3159 5833)
Also recorded as Uphill 7, this was the upper of two *'irregular and low-roofed caves'* discovered in 1898, close to the position of Uphill Quarry Cave 6. It was investigated by Edward Wilson of the British Museum, who discovered Pleistocene remains which can still be seen in the British Museum, the Bristol City Museum, and the local Woodspring Museum. Among the finds were bones of cave lion, hyaena, bear, woolly mammoth, rhinoceros and reindeer. The cave was destroyed within a few years.

Ref: E. Wilson & S. H. Reynolds, The Uphill bone caves, Proceedings of the Bristol Natural History Society 9 pp 152-160 (1901)

UPHILL QUARRY CAVE 8 (ST 3159 5833)
Also recorded as Uphill 8, this was the lower and larger of the two caves discovered in 1898. A bedding plane, floored with cave earth at least 2m deep, led to a chamber containing a number of fissures, including a chimney 2m in diameter. As well as the Pleistocene fauna associated with Uphill Quarry Cave 7, the cave also yielded a collection of Middle and Upper Palaeolithic flints, including blades, a hand axe and a spearhead of *'rude character'*. Wilson's collection also includes an antler point, which although undocumented is believed to have been recovered from this cave. This has been dated to 31,730±250 BP, and as such, represents the only clearly Aurignacian artefact discovered in the British Isles to date. Although it is believed to have been completely destroyed, the cave is thought to have been situated quite close to Uphill Quarry Cave 12, which may be a surviving remnant.

Ref: E. Wilson & S. H. Reynolds, The Uphill bone caves, Proceedings of the Bristol Natural History Society 9 pp 152-160 (1901)

UPHILL QUARRY CAVE 9 (ST 3159 5833)
Also recorded as Uphill 9, this was one of two caves examined under the auspices of the British Association between 1899 and 1901. It was described as large, and was explored for '*some distance*', but was found to be barren. Its precise location remains unknown.

Ref: C. L. Morgan et al. Caves at Uphill (1899), Ossiferous caves at Uphill (1900), Ossiferous caves at Uphill (1901), Reports of the British Association, 69 (402), 70 (342-343), 71 (352)

UPHILL QUARRY CAVE 10 (ST 3159 5833) L 7m
Also recorded as Uphill 10, this was the other cave investigated by the British Association. A short passage led to a rock shelter. Finds included gnawed bones, hammerstones, pot-boilers (round stones), and a piece of black Roman pottery. Like Uphill Quarry Cave 9, its precise location remains unknown.

Ref: C. L. Morgan et al. Caves at Uphill (1899), Ossiferous caves at Uphill (1900), Ossiferous caves at Uphill (1901), Reports of the British Association, 69 (402), 70 (342-343), 71 (352)

UPHILL QUARRY CAVE 11 (ST 3158 5836) L 6m VR 2m Alt 14m
Also recorded as Uphill 11, this surviving remnant is located several metres above the quarry floor, inside a securely fenced enclosure. There is scalloping in the cliff nearby and a number of small open bedding planes formed along a line of weakness. It features a descending bedding plane, which may represent a fragment of one of the lost caves investigated around the turn of the 20th century.

Ref: N. Barrington, W. Stanton, The Complete Caves and a view of the hills, p 167 (1977)

UPHILL QUARRY CAVE 12 (ST 3159 5833) L 11m VR 2m Alt 9m
Also secured inside the locked enclosure, and recorded as Uphill 12, this is the largest of the caves still extant in Uphill Quarry and may represent a surviving fragment of one of the lost caves investigated around the turn of the 20th century. Stanton suggested that this may be a remnant of the lost Uphill Cavern but given the horizontal separation this cannot be correct. There is another very small open hole situated directly above the entrance.

Ref: N. Barrington, W. Stanton, The Complete Caves and a view of the hills, p 167 (1977)

UPHILL QUARRY CAVE 13 (ST 3150 5844) L 4m VR 2m Alt 12m
The most westerly of all the caves discovered in the quarry, this surviving remnant is situated in a cliff, immediately above and behind the Wharf Side Marine shed. Reached by a short traverse and easy scramble, it descends westwards, narrowing rapidly.

Ref: N. Barrington, W. Stanton, The Complete Caves and a view of the hills, p 167 (1977)

Uphill Quarry. Photo by Rob Taviner

SOUTH QUARRY CAVE
Cave
ST 3164 5847   L 2m VR 1m Alt 12m

Situated diagonally opposite The Dolphin, this tiny blind alcove is located in a cliff positioned some 5m above the eastern edge of the car park.

Ref: R. Gardner, South Quarry Cave, ACG N/L (Dec 1972)

THE MYSTERIOUS UPHILL BULGE
Bulge!

In 'Britannia Baconia,' published in 1661, there is an allusion to an extraordinary incident, the location of which is quite unknown.

"It is reported that about Uphill (a parish by the sea-side not far from Axbridge) within these half-hundred years, a parcel of ground swelled up like a hill, and on a sudden clave asunder, and fell down again into the earth, and in the place of it remains a great pool."

Britannia Baconia was the most famous work of Joshua Childrey (1623-1670), an English academic, antiquary and astrologer, who was archdeacon of Salisbury from 1664. The book was derived from 100 notebooks, each devoted to one of the topics of the *Parasceve* of Francis Bacon. However, many of the descriptions of the curiosities mentioned in its pages seem little more than hearsay, a stark contrast to the more scientific methods of observation preferred by Bacon. It is unknown whether this event was in any way cave related, or even reality related, and may just have been a popular local myth of the time. Readers are advised to take a large pinch of salt and form their own conclusions.

OLDMIXON

HAY WOOD SPRING
ST 3395 5827
Spring
Alt 55m
A small, intermittent spring situated just below the track beneath Hay Wood Rock Shelter.

HAY WOOD ROCK SHELTER
ST 3400 5826
Cave
L 15m VR 3m Alt 60m
Also known as Hay Wood Cave and Hey Wood Cave, the roomy entrance to this important cave is situated in a low cliff above the track. A crawl leads to a small, moonmilk-encrusted chamber. The entrance pit was once covered with a makeshift shed, the remains of which can be seen nearby and presumably dates from 1957-1962, when excavations by the ACG & AS unearthed a quantity of Mesolithic microliths, along with sherds of Iron Age and Roman pottery. Most of the finds came from a much-disturbed mound and included several hundred animal bone fragments and more than five hundred fragments of human bone, including skulls. These were originally considered to represent Iron Age (or possibly Roman) burials, but dating later showed at least one to be Neolithic. The finds are in Axbridge Museum.

Ref: N. Barrington, W. Stanton, The Complete Caves and a view of the hills, p 94 (1977) & V. Simmonds, An overview of the archaeology of Mendip caves and karst, www.mcra.org.uk, (2014)

HUTTON HILL HOLE
ST 3430 5814
Cave
L 10m VR 6m Alt 85m
Discovered and excavated by BDCC in 1994, the roomy entrance to this cave is situated on a wooded hillside and is quite invisible from below. A diminishing tunnel descends steeply to a choke, passing a small offshoot, which can be reached with the aid of a rotting plank. Several bones, including Arctic hare were recovered.

Ref: M. Norton, An Article from Michael Norton, ACG N/L p 21 (Summer-Autumn 1995)

HAY WOOD RIFT
ST 3464 5817
Mine
L 6m VR 2m Alt 65m
This narrow open trench was dug by ACG/BDCC in 1995-96. A small natural cavity was found at one end, along with some bottles and an old iron wedge. It may have been a trial trench for ochre and has been partially backfilled.

Ref: M. Norton, Hay Wood Rift, ACG N/L p 9 (1996)

HILLSIDE COTTAGE SEEPAGE
ST 3395 5827
Seepage
Alt 55m
A boggy, bowl-shaped hollow located close to the track in the woods above Hillside Cottage.

HUTTON

Hutton is probably derived from a combination of Anglo-Saxon hoh and tun, meaning the settlement on the spur. Several important caves have been discovered on this spur (Hutton Hill), including Bleadon Cavern and Upper Canada Cave, which are now known to be the modern counterparts of two celebrated, but long-lost antediluvian sites - both confusingly called Hutton Cave!

LUDWELL CAVE
Cave / Rising
ST 3590 5924
L 40m VR 6m Alt 16m

This small cave, located partway between Hutton and Locking, comprises a low passage leading into a rift chamber, which is formed along the junction of the Dolomitic Conglomerate and Carboniferous limestone. A stream flows into a sump pool and reappears almost immediately at **Ludwell Rising**, which is located just below the cave entrance. The water level in the cave rises by almost a metre during very wet weather and while the source of the stream remains unknown, it is probably fed by water penetrating into a spur of limestone extending to the south-east. A local legend which holds that the water could once be clearly heard from Locking church, suggests that the spring was once significantly larger, and it is said to have further diminished in size following the sinking of a borehole at Locking's Bristol Aerojet factory in 1940. The cave was first entered in 1951, when Willie Stanton free-dived from the resurgence to the sump pool and small chamber. In 1972, Mike 'Fish' Jeanmaire of CDG recorded negotiating an underwater boulder choke and entering some 30m of large, underwater passage. However, the boulder choke appears to move during flood conditions and the reported passage has not been seen or entered since. Ludwell (or Lodwell) may derive in part from Lludd, a Celtic deity. However, several other springs and holy wells carry this name, and it is much more likely to have derived from Olde English 'hlud' and 'waella', meaning loud or strong spring.

Ref: Mendip Underground, MCRA p 209 (2013)

CANADA COMBE

CANADA COMBE CAVE
Cave
ST 3604 5852
L 14m VR 5m Alt 61m

Opened by ACG in 1995 and situated on the steep western flank of Canada Combe some 8m above the road, this small phreatic network has two entrances and contains a notable eroded calcite vein. In 1965, nearby **Canada Combe Fissure** (ST 3605 5840) was also dug briefly by ACG, before being backfilled. No clear sign of this earlier fissure now remains and given the proximity of the two NGRs it may be that the two sites were actually one and the same.

Ref: M. Norton, Canada Combe Cave A.A.H (After Albanian Hilti), ACG Jnl 74 pp 2-3 (1998)

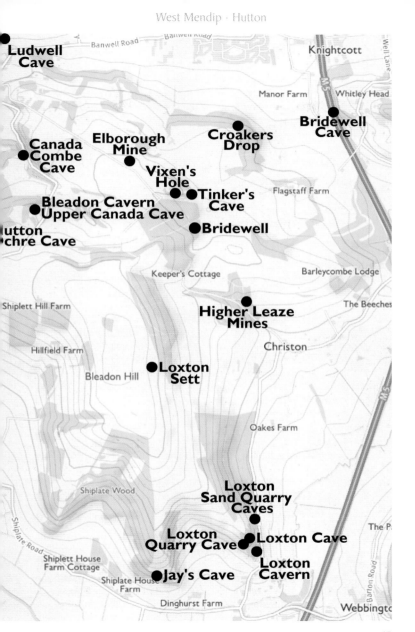

CANADA COMBE ROCK SHELTER — Cave
ST 3616 5819 L 3m VR 2m Alt 97m

Situated directly opposite Canada Farm, this cave is little more than a small overhang at the head of a small rocky gulley. In 1971, ACG excavated a tiny tube, which was heading in the general direction of the (then) Lost Cave of Hutton. Local tradition holds that foxes sometimes hide here, to evade the hunt – hence the alternative name of Foxes Hole.

Ref: N. Barrington, W. Stanton, The Complete Caves and a view of the hills, p 48 (1977)

HUTTON HILL
ANTEDILUVIAN CAVES

The fields and woods above Upper Canada once yielded significant deposits of both yellow ochre and calamine, and it was here, c.1746, that William Glisson and his ochre miners, discovered a cavern containing numerous ancient animal bones. Some of these were extracted by Dr Alexander Catcott (a noted antediluvian) in 1757, who presented them to Bristol Library Museum. Sadly, this important assemblage was later destroyed during a WWII bombing raid. The original discovery was described by Rutter (1829) as follows.

'The miners having opened an ochre-pit, came to a fissure in the limestone rock filled with good ochre, which, being continued to the depth of eight yards, opened into a cavern, the floor of which consisted also of ochre ; and strewed on its surface were large quantities of white bones, which were found dispersed through the ochreous mass. In the centre of the chamber, a large stalactite depended from the roof, beneath which a corresponding pillar of stalagmite arose from the floor.'

Dr Catcott *'descended into a cavern about ninety feet deep, around whose sides, and from the roof, the bones projected, so as to represent the inside of a charnel-house."*

Shortly afterwards, the walls fell in, and the cavern was lost from view. Then in 1828, the Rev. David Williams found some fragments of bone among the rubbish near the mouth of an old pit, and started sinking shafts, the third of which intercepted a natural cave system. With the assistance of William Beard, Williams made a thorough exploration of the place, which seems to have been a fissure filled with ochreous rubble, in which were found yet more great quantities of Pleistocene animal bones, including those of the elephant, lion, hyaena, wolf, boar, and horse. In 1833, Beard's men opened yet another cave, which Beard initially called his '*Second Cavern of Bones*'. To aid with the excavation Beard commissioned a survey, which remarkably survives intact and can be viewed in the Somerset Heritage Centre. Information extracted from this ochre-stained drawing proved crucial in untangling the long-standing mystery surrounding the Lost Cave of Hutton. This 'mystery' was due largely to Rutter, who in 1829, published both Catcott's original account, and Williams later account, together, thus giving the distinct impression that these were two accounts of the same cave. In the 1970s, cavers' re-excavated a number of old shafts in an attempt to relocate it and many interesting discoveries were made, including Bleadon Cavern, which was quickly identified as the later cavern, opened by Beard in 1833. However, none appeared to match the description of the cave published by Rutter in 1829. Further excavations in the same area by ACG, beginning in 2006, gradually uncovered the secrets of Upper Canada Cave, and the combination of this, and the earlier work, ultimately led to the realisation that Catcott's lost cave, and the one subsequently explored by Williams and Beard, were not, in fact, the same cave after all, but were actually two entirely different caves, both coincidentally bearing important bone assemblages. Catcott's Hutton Cavern has now been positively identified as Bleadon Cavern, while the 1828 cave explored by Williams and Beard satisfactorily matches the upper reaches of Upper Canada Cave. Just to add to the confusion, Beard's *Second Cavern of Bones* is also Bleadon Cavern. To complicate matters even further, other references to a 'Lost Cavern of Hutton' give a much earlier date of c.1650! This date however, is derived from F. A. Knight's, Seaboard of Mendip, who appears to have muddled the work carried out by Williams and Beard, with discoveries elsewhere.

Ref: J. Rutter, Delineations of the North Western Division of the County of Somerset, Longman Rees & Co. pp 100-104 (1829), D. J. Irwin & C. Richards, The Bleadon and Hutton Caverns, West Mendip, BEC Belfry Bull 496 pp 38-49 (1998) & N. Richards & N. Harding. The Search for Hutton Cavern, BEC Belfry Bull 524 pp 29-41 (2006)

HUTTON CAVERN (Catcott variant 1757)   Cave

Also known as Hutton Cavern - I, this is the famous, original 'lost' cave first opened by the ochre miners sometime between 1739 and 1746 and visited and assiduously recorded by Catcott in 1757. He wrote several letters describing the cave and one of the best was reported in *The Annals of Philosophy*, and also by William Buckland in his *Reliquiae diluvianae* (1823). We are fortunate indeed that such a detailed description exists, which would have done credit to Mendip Underground!

'The ochre was pursued through fissures in the mountain limestone, occasionally expanding into larger cavernous chambers, their range being in steep descent, and almost perpendicular. This, in opening the pits, the workmen, after removing 18 inches of vegetable mould, and four feet of rubbly ochre, came to a fissure in the limestone rock, about 18 inches broad, and four feet long. This was filled with good ochre, but as yet no bones were discovered; it continued to the depth of eight yards, and then opened into a cavern about 20 feet square, and four high; the floor of this cave consisted of good ochre strewed on the surface of which were multitudes of white bones, which were also found dispersed through the interior of the ochreous mass. In the centre of this chamber, a large stalactite depended from the roof; and beneath, a

Bleadon Cavern. Smoke mark dated 1746. Photo by Alan Gray

similar mass rose from the floor, almost touching it: in one of the side walls was an opening about three feet square, which conducted through a passage 18 yards in length to a second cavern 10 yards in length and five in breadth, both the passage and cavern being filled with ochre and bones; another passage, about six feet square, branched off laterally from this chamber about four yards below its entrance; this continued nearly on the same level for 18 yards; it was filled with rubbly ochre, fragments of limestone rounded by attrition, and lead ore confusedly mixed together; many large bones occurring in the mass; among which four magnificent teeth of an elephant (the whole number belonging to a single skull) were found; another shaft was sunk from the surface perpendicularly into this branch, and appears to have followed the course of a fissure, since it is said that all the way nothing appeared but rubble, large stones, ochre and bones: in the second chamber, immediately beyond the entrance of the branch just described, there appeared a large deep opening, tending perpendicularly downwards, filled with the same congeries of rubble, ochre, bones etc; this was cleared to a depth of five yards; this point being the deepest part of the workings, was estimated at about 36 yards beneath the surface of the hill; a few yards to the west of this a similar hole occurred, in which was found a large head, which we shall have occasion presently to notice.'

Catcott's diaries said that the entrance lay 'three hundred paces south of the gate into Down Acres'. Glisson, the chief ochre miner, claimed that he had opened one large bone cave and several smaller ones, and this evidence, coupled with the accuracy of the description and a smoke mark dated 1746 on the wall in Bleadon Cavern (which proves that the cave had been opened prior to Catcott's visit), eventually led to the realisation that the modern Bleadon Cavern, and Catcott's Hutton Cavern, are in fact one and the same. Catcott's original entrance however was located on the other side of the wall in Bleadon parish and this fits well with F. A. Knight's record that the mouth of the cavern was still visible *'in the centre of a group of Scotch firs, in a field adjoining the road, not far above Upper Canada Farm'*. This seems to have been confirmed by Stanton in 1944, who described visiting a 6-feet square hole *'8 feet deep'*, while old cave registry sheets record a choked 3m deep mineshaft (Fir Tree Mine), in much the same location.

Ref: F. A. Knight, The Sea-Board of Mendip, J. M. Dent & Company pp 377-378 (1902) & W. I. Stanton, Logbook 2 p 3 (1944)

HUTTON CAVERN 2   Cave

Catcott recorded a second cavern opened by the 18th century ochre miners, which he called the lesser cave. The cave was apparently located about 40m west of his Hutton Cavern, and until recently was thought lost. One miner, who brought Catcott a selection of small bones, described finding the head of a strange animal, *'about 3 or 4 feet long: 14 inches broad at the top or hind part and 3 inches at the snout shaped like a crocodile'*. The skull apparently had four tusks, two on each jaw and with large teeth, which Catcott associated with a hippopotamus. The miner claimed that he'd hidden the skull in a nearby wood, but when he went back later with Catcott to reclaim it, he was unable to find the spot. The skull was probably that of a boar. This is now known to be the cave excavated by Williams and Beard in 1828, represented today by the upper reaches of Upper Canada Cave.

HUTTON CAVERN (Williams and Beard variant 1828)   Cave

In 1828, the Rev. David Williams began searching for Catcott's lost cave and discovered some fragments of bone among the rubbish near the mouth of an old pit, which had been pointed out to him by the son of an old miner. He began sinking shafts (represented today by Archway Cave, May Tree Cave and Primrose Cave), the third of which (Primrose Cave) intercepted a natural fissure filled with bones and ochreous rubble. Great quantities of Pleistocene animal bones, subsequently dated to 190,000 BP were recovered, including those of numerous juvenile horses, possibly deposited by a flood or other debris flow event. They described the cave as *'three chambers in the fissure, the floor of the one above, forming the roof of the one below, and consisting of huge fragments of rock, which have sunk away and jammed themselves between the strata; their intersections being filled with ochreous rubble and bones'*. Williams even drew a famous elevation of the cave, which appeared alongside Catcott's description, in Rutter's Delineations of the North Western Division of the County of Somerset. Until recently everyone assumed this to be a remnant of Catcott's famous lost cavern, but recent investigations have shown that the two caves were entirely different, and that Williams had actually re-

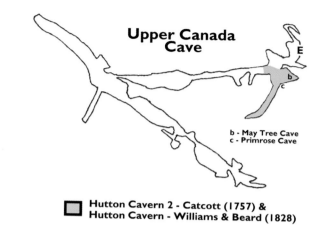

**Hutton Cavern 2 - Catcott (1757) &
Hutton Cavern - Williams & Beard (1828)**

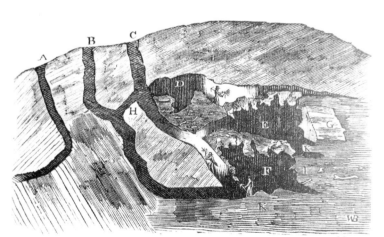

entered Catcott's lesser cave, Hutton Cavern 2, which is now known to be the upper reaches of Upper Canada Cave. Williams himself never seems to claim that he had rediscovered Catcott's main cave – recording only that Catcott had earlier recovered bones from '*this hill*'. In fact, Williams, writing in 1834, actually refers to the site as 'Bleadon Cave' (not to be confused with the modern Bleadon Cavern!), which suggests that he was well aware of the distinction.

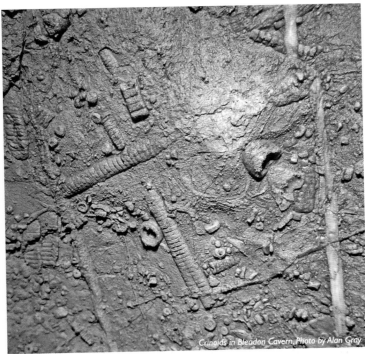

Crinoids in Bleadon Cavern. Photo by Alan Gray

who recorded it as the Second Cavern of Bones on Hutton Hill. Initially entered from the Hutton side of the parish boundary, Beard's men later sank a second entrance in neighbouring Bleadon parish (probably using Catcott's original entrance), to facilitate removal of the bones, after which time the site was lost until ACG reopened Beard's 'Hutton parish' entrance in 1970. The cave offers much geological and archaeological interest combined with some sporting climbs and squeezes. There are drag-rope marks, walls of 'deads', trenches, wooden roof supports and spoil dumps throughout, plus some extensive bone deposits, sadly much-ransacked in the early 19thcentury. The bones, which make up one of the most important assemblages in southern England, have been dated to 250,000 BP, and the remains of several adult lions suggest that the cave was once a large cat den. Undated human bones have also been found in the material. Old maps of the area clearly refer to this site as Hutton Cavern, and this, along with the smoke mark evidence, and notes annotated to the original dialling survey, prove beyond doubt that this is the original 'lost' bone cave explored by Catcott in 1757.

Ref: D. J. Irwin & C. Richards, The Bleadon and Hutton Caverns, West Mendip, BEC Belfry Bull 496 pp 38-49 (1998) & Mendip Underground, MCRA pp 66-68 (2013)

COLE CHUTE
Cave
ST 3605 5815 L 28m VR 8m Alt 107m
Situated halfway between Bleadon Cavern and the Upper Canada Cave complex, this small cave comprises two entrances (covered with farm gates) which yield an unstable chamber and a complex of crawls. Originally worked by ochre miners, it was re-opened by BEC in 2003 and again by ACG in 2017 who discovered a miner's wedge embedded in an ochre-filled crevice. The cave is named after the former landowner Bernard Cole, who sadly passed away in 2017.

Ref: N. Richards & N. Harding, More Pits Than You Can Shake A Stick At, BEC Belfry Bull 530 p 31 (2008)

GALLERY PIT CAVE
Cave
ST 3602 5815 L 20m VR 10m Alt 108m
Located on the upper bench immediately opposite the Primrose Cave entrance to Upper Canada Cave, this site comprises a loose crawl to a descending bedding plane containing stacked deads. It was opened by BEC in 2006 and sealed shortly after.

Ref: N. Richards & N. Harding, Hutton Discoveries – Gallery Pit Cave, BEC Belfry Bull 526 pp 27-32 (2006)

ARCHWAY CAVE
Mine
ST 3603 5816 L 11m VR 11m Alt 108m
This open chimney is best accessed via the open pit directly opposite the entrance to Upper Canada Cave. It leads to a steeply-descending mined passage which represents the unsuccessful mineshaft depicted on the famous elevation published by Rutter in 1829, and drawn, according to Williams, to a depth of '*70 feet*'. The surface opening was once known as Draughting Pit, but both this, and several neighbouring shafts (Blind Pit, Two Trees Pit and Schrapnel Hole), were swallowed up when the large open pit was excavated during 2012.

Ref: J. Rutter, Delineations of the North Western Division of the County of Somerset, Longman Rees & Co. pp 100-104 (1829)

UPPER CANADA CAVE
Cave
ST 3603 5816 L 157m VR 29m Alt 108m
Located in an area of intensive 18th century ochre workings, Upper Canada Cave has several entrances surrounding a broad open pit. It comprises a series of quite roomy, sloping chambers, connected by narrow rifts and constricted ochreous crawls. The upper reaches of the cave are an amalgamation of Primrose Cave and May Tree Cave - two caves originally explored by ACG during the 1970s. These were linked together in 2012 following work with a mechanical digger. The lower, main part of the cave was originally entered by Nick Harding and Nick Richards in 2007 during their intensive search for the famous lost Hutton Cavern. Hundreds of bones have been recovered, many of which proved to be almost entirely modern, doubtless the remains of carcasses dumped down the shafts. However, at least one bone, that of a Pleistocene horse, has recently been radiocarbon dated to 21170 +/- 70 BP. The upper reaches of Upper Canada

Upper Canada Cave. Photo by Steve Sharp

Cave is the site excavated by Williams and Beard in 1828, as evidenced by the initials 'DW' smoke-marked onto the wall in the Primrose Cave. This is almost certainly the lesser bone cave (see Hutton Cavern 2) recorded by Catcott in 1757, which was reopened by ACG and Hutton Scouts in 1973/74, who called it Hutton Cavern 3. The cave was originally entered via a vertical shaft, but this was one of several that vanished (Blind Pit, Two Trees Pit, Draughting Pit and Schrapnel Hole), when the large open pit was excavated during 2012.

Ref: D. J. Irwin & C. Richards, The Bleadon and Hutton Caverns, West Mendip, BEC Belfry Bull 496 pp 38-49 (1998) & Mendip Underground, MCRA pp 421-422 (2013)

WELL SHAFT CAVE

Cave

ST 3603 5817

L 31m VR 11m Alt 115m

At the extreme north-east corner of the deep pit containing Upper Canada Cave, is a steep cliff displaying remarkable fossil algal stromatolites. At its base a tiny hole yields a low bedding plane, which descends steeply to reach the Well Shaft, a 5m deep, ginged pot which terminates in a blind ochre pocket. Across the shaft, a passage leads to a climb up a rift which emerges in the depression immediately adjacent to the entrance to Upper Canada Cave. The cave contains a smoke mark dated 1768, left by William Glisson's sons, and there is another loose passage which leads to an impassable connection with 'The Great Wall' in Upper Canada Cave. The cave was originally opened by ACG in 1973, who named it Hutton Cavern 4. It was quickly refilled but was reopened during work to rediscover the lost 'Hutton Cavern' in 2012.

Ref: D. J. Irwin & C. Richards, The Bleadon and Hutton Caverns, West Mendip, BEC Belfry Bull 496 pp 38-49 (1998) & Mendip Underground, MCRA p 421 (2013)

UNDER SHED PIT (ST 3603 5812) – There is a large shed within the Bleadon Cavern compound, one corner of which overlies loose ground. This is probably the entrance to an old mine where an unseen hole reportedly swallowed a sizeable amount of material. Steel beams were required to help stabilise the structure. There are a great many other ochre pits nearby, several of which were recorded, dug and backfilled in the 1970s and 2000s, during the hunt for the 'Lost Cave of Hutton'. Some (**Blind Pit, Two Trees Pit, Draughting Pit** and **Schrapnel Hole**), were swallowed up when the large open pit was excavated during 2012, while others, including **Tyre Pit, Tree Stump Shaft** (ST 3604 5815), **Fence Pit, Small Pit, Hutton Shaft 1** (ST 3607 5813) and **Hutton Shaft 2** (ST 3610 5814), were noted but never amounted to much. The latter two were recorded as capped mineshafts of unknown depth.

BEDROCK PIT (ST 3600 5818) L 3m VR 3m Alt 104m - Originally named Croc Pit in the mistaken belief that it might be the site of Catcott's lost Hutton Cavern 2, this hole was excavated down to bedrock by ACG in 2014-2015.

HUTTON WOOD

HUTTON CAVERN DIG 1
ST 3592 5815　　　　　　　　　　　　　　　　　　　　　Cave
　　　　　　　　　　　　　　　　　　　　　L 9m VR 5m Alt 105m

In 2005, during their search for the Lost Cave of Hutton, BEC sank three shafts into an unroofed phreatic chamber which had been backfilled with miners' rubble. A clay pipe dating from 1820-1840 was found. The site has been completely backfilled and is now barely discernible from any of the many other overgrown hollows in this part of the wood.

Ref: N. Richards & N. Harding. The Search for Hutton Cavern, BEC Belfry Bull 524 pp 29-41 (2006)

HUTTON CAVERN DIG 2
Mine
ST 3589 5815   L 17m VR 5m Alt 105m

This old ochre mine was another excavated by the BEC in 2006. A shaft yielded a rift and a horizontal stal-lined tunnel some 10m long. A trial dig in the adjacent pit revealed little and the mine has since been thoroughly backfilled.

Ref: N. Richards & N. Harding. The Search for Hutton Cavern, BEC Belfry Bull 524 pp 29-41 (2006)

HUTTON OCHRE CAVE
Mine
ST 3585 5807   L 18m VR 6m Alt 116m

Also known as Hutton Wood Mine, this isolated working comprised short crawls with choked connections to the surface and adjacent pits. It was excavated by '*The Two Nick's*' c.2006, who found clay pipes in wall niches. The pit has been partially filled with brushwood and the entrance has collapsed. Digs in the adjacent pits hit bedrock at -2m.

Ref: N. Richards, Personal communication

TUMBLEDOWN WALL HOLE
Mine
ST 3582 5810   L 1m VR 1m Alt 45m

Located close to the tumbledown wall which snakes its way through the eastern part of the wood, this tiny ochre-filled hole, situated below a small cliff, is actually the top of a rubble-filled mineshaft. It was dug briefly by '*The Two Nick's*' c.2006.

Ref: N. Richards, Personal communication

GLEBE WOOD SPRING
Spring
ST 3572 5839   Alt 45m

Located just below the track on the northern edge of Hutton Wood, this small marshy gulley emits a lively stream in wet weather. It was dug briefly by ACG in 2016, who named it Twyford Dig, after a small sink located 30m further west. It is situated directly above Ladies Well and may act as a flood overflow.

LADIES WELL
Spring
ST 3558 5852   Alt 30m

Local folklore holds that this spring, which rises from a small stone 'dolmen', once fed a bathing house used by both nuns and monks stationed nearby, possibly during the period when Hutton Court was owned by John Still, the Bishop of Bath and Wells. Also recorded as Lady's Well or Ladywell, it was more likely utilised to fill the court's fish ponds. Hutton was once almost entirely dependent on water extracted from wells, notably the **School House Well** (ST 3522 5876), which was the village's main water supply.

ELBOROUGH

ELBOROUGH MINE   Mine
ST 3672 5843   Alt 110m

Originally operated by local men, this mine, which ultimately yielded barytes, lead, calamine and ochre, was taken over in 1846 by Cornish miners acting under the control of Richard Trevithick. Several shafts were sunk, including Bickford's, Chapman's and Vivian's, which operated on different lodes. Vivian's Shaft was said to be 33m deep in 1856, and a large natural cavern was apparently encountered at -29m. The NGR is centred on a spot where numerous capped shafts are evident.

Ref: M. Clarke, N. Gregory and A. Gray, Earth Colours – Mendip and Bristol Ochre Mining, MCRA, p 155 (2012)

VIXEN'S HOLE   Cave
ST 3699 5822   L 5m VR 2m Alt 100m

Mired in brambles and best approached from above, this cave, which is also known as Foxes Cave, consists of a wide, very low bedding plane, clearly much favoured by animals. Located on the southern slopes of Elborough hillfort, it was pushed by Hutton Scouts in 1976 to a positive conclusion but has now partially refilled. This may be a 'lost cave' that Charles Ponsford, a former resident of Canada Combe, claimed ' ... *which goes back under*.' Tinker's Cave is another possibility.

Ref: N. Barrington, W. Stanton, The Complete Caves and a view of the hills, p 168 (1977)

TINKER'S CAVE   Cave
ST 3707 5821   L 12m VR 6m Alt 85m

Located at the south-east corner of Elborough hillfort, and also known as Tinker's Hole, this cave was excavated by ACG in 1963-66. A descending tunnel yields a narrow twisting passage.

Ref: N. Barrington, W. Stanton, The Complete Caves and a view of the hills, p 162 (1977)

CROAKERS DROP   Mine
ST 3736 5866   L 4m VR 4m Alt 65m

This narrow open shaft, located close to the southern edge of Benthills Wood, was first noted by ACG in 1996. There is a large open pit a short distance further north.

Ref: ACG Logbook No.8 pp 52-53 (1996)

BRIDEWELL   Spring
ST 3710 5801   Alt 89m

Located in woodland at the foot of a low scarp, adjacent to an overgrown tank capped with concrete, this small, shallow, inauspicious boggy hollow is the suggested site for the 'lost' well from which Bridewell Lane derives its name. In wet conditions, a wide body of water appears and pours out onto the track, which was presumably constructed to connect Banwell with the water source. Like many similar named wells, Bridewell is probably a corruption of St. Bridget's Well, which in turn may be a Christian version of an earlier pagan name. Bridewell Lane also lies alongside the route of a proposed Roman road linking Charterhouse with a possible port at Uphill. Unusually, the site of the well is not marked on old maps, which suggests that it went out of regular use a long time ago. A short distance to the north there is an active pump house which is said to have affected the flow at Ludwell Rising.

Ref: N. Richards, Personal communication

HIGHER LEAZE MINE 1 — Mine
ST 3742 5742 L 3m VR 3m Alt 69m

The most northerly of a tightly-grouped cluster of old lead and calamine workings located on the northern slopes of Higher Leaze, this shaft, which is currently blocked with wire (2018), enters a small working heading in the direction of Higher Leaze Mine 2.

HIGHER LEAZE MINE 2 — Mine
ST 3742 5751 L 12m VR 4m Alt 68m

The most impressive of the surviving Higher Leaze mines, a steep chute, floored with glass, enters a roomy rubbish-strewn chamber. The continuation leading towards Higher Leaze Mine 1 is choked.

HIGHER LEAZE MINE 3 — Mine
ST 3743 5751 L 9m VR 4m Alt 69m

An open rift, with a low connection to Higher Leaze Mine 4 which is currently blocked (2018).

HIGHER LEAZE MINE 4 — Mine
ST 3743 5750 L 6m VR 4m Alt 68m

A narrow rift descends to an unroofed chamber, where there is a low connection to Higher Leaze Mine 3 which is currently blocked (2018). There are a number of choked shafts in the gruffy ground to the south-east.

CHRISTON SPRING — Spring
ST 3830 5722 Alt 12m

This small, intermittent spring is the feeder for an inlet into the Lox Yeo River. It used to flow under the Ox House (an alternative name), a building which was demolished during the construction of the M5. There is another interesting water source in **Barleycombe Wood** (ST 3820 5789 Alt 26m), which may have supplied water to nearby ancient settlements.

BLEADON

Bleadon is derived from *Bleodun*, meaning the 'coloured hill', and the hill contains a 13.5 hectare geological SSSI. It is also classified as a Geological Conservation Review site, as it shows a low ridge of calcite-cemented Pleistocene sand and gravel, which probably accumulated during the Quaternary, when an ice sheet rested against the side of Bleadon Hill.

PURN HILL CAVE
Cave
ST 3322 5708
L 11m VR 3m Alt 12m

Also known as Purn Hill Mine, this isolated cave is situated on an overgrown bank at the southern point of Purn Hill, a 6.1 hectare SSSI, cited for its exceptionally diverse unimproved calcareous grassland flora. It is located directly opposite the restaurant adjacent to the Anchor Inn. The entrance, which is obscured by ivy, yields a relatively roomy entrance tunnel, which is clearly used by adventurous local youths as an illicit drinking den! There is some unusual brown staining on the walls and the cave terminates soon after in a diminishing crawl. There are several insignificant fissures in the quarried breccia cliffs at road level.

Ref: N. Barrington, W. Stanton, The Complete Caves and a view of the hills, p 125 (1977)

PURN HILL MINES

The southern end of the summit of Purn Hill is dotted with mined pits and recent scrub clearance revealed two very small mines near the top of the eastern slope. **Purn Hill Mine 1** (ST 3324 5721, L 3m Alt 45m) consists of a sloping passage in solid rock, favoured by badgers, while **Purn Hill Mine 2** (ST 3324 5722, L 2m Alt 45m) is a small tube which rapidly chokes. Both mines are now obscured by debris.

Ref: N. Richards, Personal communication

BLEADON QUARRY (LITTLE DOWN QUARRY)

Bleadon Quarry began life in the late 19th century as Little Down Quarry. Several small but notable caves have been encountered, which appear to be connected by at least one underground streamway crossing the quarry from north-east to south-west. One contained important bone deposits, which were extracted by Wadham Pigott Williams (the quarry owner and eldest son of Rev. David Williams) working alongside a Mr. Bidgood. The bones, which dated from the Pleistocene period, included wolf, tiger, ox, horse, bear and elephant, with the bear and elephant in particular reportedly being of prodigious size.

BLEADON QUARRY CAVE 1
Cave
ST 3418 5674

This was the first cave discovered in the quarry, which was described in some detail by B. Cox of Lympsham in a letter to the Weston-super-Mare Gazette dated Aug 12th 1854.

'In the first place, then, you enter a hole formed in the rock, some eight feet high, by five feet wide, and about fifteen feet long, descending nearly ten feet from a level from the entrance; in this passage there is no particular beauty to attract the notice of its visitor except the irregularity of its rocky sides, and its hard incrustation. After reaching the end of this passage, you arrive at a large perpendicular cave, fourteen feet over [wide], which you descend by means of a ladder to the depth of nearly twenty feet, somewhat of a Gothic design; you traverse this cavern in a south and west direction until you come to a large body of pure water on each side, its abyss being quite unfathomable, and the end cannot, as yet, be ascertained, for if a lighted candle be fastened on a floating vessel, it will gradually draw off, until it is gone completely out of sight. When this cavern was first opened it was a beautiful and splendid sight to behold its various stalactites hanging from the roof formed by an accumulation of those little concretions in the drops of water, which, falling from the roof, have, by a long period of time, acquired the hardness of marble; those beautiful solid icicles, transparent as glass, have been broken and scattered in every direction by the rude hands of many of its visitors…'

The cave, which is marked by a pile of boulders located beneath the quarry face in the overgrown northern embayment of the quarry, was visited by Mr. Merchant, the Chairman of Burnham Local Board of Health in 1884, who found a quantity of water *rambling around the stones*. Presumably he was trying to ascertain whether it could supply a sufficient water supply for his town, although this plan was clearly abandoned in favour of the development of Cox's Spring in Winscombe. The cave was apparently blocked with rubble in 1919 to keep inquisitive children out.

BLEADON BONE CAVE   Cave
ST 3418 5675 approx

Also known as Bleadon Cave, this was perhaps the most important cave discovered in the quarry. The bones, at least some of which Williams donated to SANHS, were found *in a fissure in the mountain limestone in a quarry at Bleadon*, which he worked from 1878 to 1881. William Bidgood, the curator of Taunton Museum, who worked alongside Williams in 1881, recorded bones of bison, ox, reindeer, bear, wolf, fox, reindeer and birds from the mouth of a downward trending cave. Some of these bones can be seen in the museum collections at Weston-super-Mare, Wells and Taunton. The cave, which probably formed an upper part of Bleadon Quarry Cave 1, was subsequently quarried away.

BLEADON QUARRY WELL   Cave
ST 3407 5666

This was one of several wells sunk c.1898 during a venture to extract water from the 'spring' flowing beneath the quarry, ostensibly to supply water to the town of Highbridge. It intersected a water-filled cave, which ran north-east towards Bleadon Quarry Cave 1 and south-west towards the cottages on

Bridge Road (where several wells *were not diminished without rain*'). The supply ultimately proved insufficient although the water was later used in the production of concrete products. Highbridge eventually took its supply from the reservoir in Cheddar Gorge.

Ref: S. L. Hobbs & P. L. Smart, Delineation of the Banwell Spring Catchment Area and the nature of the spring hydrograph, UBSS Proc 18 (3) pp 359-366 (1989)

OTHER BLEADON QUARRY CAVES

- Several other cavities have been uncovered over the years, including a '*huge cavern with stalactites!*' said to have been broken into during construction of a toilet block c.1998. It is believed to have been explored by an employee although no further details are known. Located behind an old pigsty and close to both the Bone Cave and Bleadon Quarry Cave 1, **Bleadon Quarry Caves 2** and **3** (ST 3418 5675) are two small phreatic recesses. A swan was reported to have nested inside one in 1984! **Bleadon Quarry Cave 4** (ST 3403 5661) lies buried beneath concrete in the south-west extremity of the quarry. Intercepted c.1920, water could be heard flowing below the entrance, which presumably formed a downstream part of the water-bearing fissure crossing the quarry. Two small phreatic holes (**Twin Tubes**, ST 3422 5663), are visible close to the south-east corner, and further caves - **Bleadon Quarry Cave 5** (ST 3409 5670) and **Bleadon Quarry Cave 6** (ST 3412 5661), have been noted, although no details are known.

Ref: N. Richards, Personal communication

BLEADON SPRINGS

- Three springs rise in the east bank of 'New Rhyne' which runs alongside Bridge Road. The largest spring, **Bleadon Spring 1** (ST 3398 5672), rises opposite Victoria Cottage and may represent the outlet for the water discovered in Bleadon Quarry Cave 1 c.1854. **Bleadon Spring 2** (ST 3398 5659) is much smaller but probably relates to the water heard flowing beneath the south-west corner of the quarry in 1920. It rises opposite Greenwood Cottage, and it was in this vicinity that wells '*not diminished without rain*' were recorded. While it seems likely that these springs have been artificially piped to their present locations (they do not appear on old maps), the presence of more than one outfall suggests that there may be more than one water-filled fissure running under the quarry floor. A third lively spring, **Bleadon Spring 3** (ST 3397 5683), located opposite the junction between Bridge Road and Coronation Road, does not appear to be related.

SUNNY COVE HOLE
ST 3404 5656

Cave

L 2m VR 1m Alt 10m

Situated in the north face of a small quarry to the south of the main working, this small phreatic bell is located immediately behind the Sunny Cove bungalow.

Ref: N. Richards, Personal communication

HILLGATE HOUSE HOLE   Cave
ST 3407 5652   L 2m VR 1m Alt 10m

This attractive phreatic hole is situated in the south-east corner of the small quarry immediately behind Hillgate House. The similarity in height and appearance between this and Sunny Cove Hole suggests that they may once have formed part of a single truncated passage.

HELLENGE HILL HOLE   Mine
ST 3465 5725   L 6m VR 3m Alt 92m

Nothing now remains to be seen of this short, choked tunnel which was entered via a collapse in the roof. It was almost certainly a lead mine and the presence of several small overgrown mounds of tipped rubble nearby suggest that similar collapses may have occurred.

Ref: N. Barrington, W. Stanton, The Complete Caves and a view of the hills, p 95 (1977)

LOXTON

Occupied since Norman times, the village is mentioned in the Domesday Book as *Lochestone*. There was probably a great deal of early mining on the hill, for calamine, lead, copper and yellow ochre, but systematic work didn't begin until c.1757, when local man William Glisson - the man responsible for finding both Loxton Cavern, and the fabled Hutton Cavern - began directing operations. Cornishmen associated with the copper mines at Dodington in the Quantock Hills later took up the cudgel, but it was soon realised that there was actually very little lead or copper to be had, and even the local calamine was of a low quality. By 1797, the Cornishmen's venture had ceased. In 1954, the village came to national attention when Miss Noreen O'Connor, a nurse, killed Friederika Alwine Maria Buls, by 'plucking out' her eyes. She was found guilty of murder, but not surprisingly was declared insane.

SHIPLATE SLAIT

SHIPLATE SPRING   Spring
ST 3585 5630   Alt 10m

Situated in a small valley within the precincts of Shiplate Manor Farm, this spring-fed pond is mentioned as a boundary mark in an Anglo-Saxon Charter of AD 956.

JAY'S CAVE   Cave
ST 3686 5572   L 34m VR 9m Alt 30m

Situated in an overgrown depression 30m south of the track which runs across the hill (below two horse troughs), this is a natural cave excavated by ochre miners. A narrow rift enters a small chamber where a flat-out crawl yields a much larger chamber (20m x 5m x 5m). Pick marks and deads are evident throughout, and an old miner's lamp was found in situ. The cave was re-entered by ACG in 1999 following a tip off from the landowner. There is a choked fissure (**Shiplate Fissure**, ST 3680 5593) in the woods to the north, along with a few infilled red ochre pits. Shiplate Wood consists of two distinctive spurs and given their similarity to comparable landforms at Compton Bishop, Loxton and Bleadon (all of which contain significant caves), it seems likely that further small, undiscovered cave systems may be present.

Ref: D. Harris, Jays Cave - A New Dig for ACG on Western Mendip, ACG Jnl pp 38-43 (Dec 1999)

LOXTON HILL

LOXTON QUARRY CAVE
ST 3738 5593

Cave

L 45m VR 12m Alt 41m

Also known as Loxton Quarry Cave 1, the entrance is located in the south-west corner of the quarry and is fitted with a large gate. It consists of a roomy phreatic tunnel which gradually splits and degenerates. There are several features of interest, including a bone choke, several old iron tools and some pieces of pottery. One small aven ascends so close to the neighbouring house that the television has been plainly heard! The majority of bones recovered are from a pig, which may be Neolithic in origin. The furthest reaches were opened up by ACG/BDCC in 1996-1998.

Ref: J. H. Tucker, Loxton Quarry Cave, BEC Caving Rep 9 pp 14-15 (1962), M. Norton, The New Loxton Quarry Cave Extensions, ACG Jnl pp 2-3 (May 2000) & A. Gray, Cave Surveys, ACG Jnl p 33 (Mar 2013)

LOXTON QUARRY CAVE 2
ST 3739 5594

Cave

L 1m VR 1m Alt 47m

This obvious alcove is located 6m above the quarry floor opposite Loxton Quarry Cave.

LOXTON QUARRY CAVE 3
ST 3740 5593

Cave

Alt 41m

A prominent bedding plane in the eastern face of the quarry marks the site of a tight descending tunnel which was covered by infill c.1955.

Ref: N. Barrington, W. Stanton, The Complete Caves and a view of the hills, p 112 (1977)

LOXTON CAVERN
ST 3744 5589

Cave

L 277m VR 24m Alt 35m

The locked manhole entrance to this cave is located in a small quarry sandwiched between the West Mendip Way and the rear of several council houses. The fabled 'Lost Cave of Loxton', it consists of a series of once well-decorated chambers developed at several levels along an east-west alignment. The cave contains many smashed formations and several notable points of interest, including a clay pipe, and 18th

Loxton Cavern. Photo by Steve Sharp

century candle smoke graffiti and knee prints. It was first entered by William Glisson's ochre miners in 1757 and described in detail soon after by the antiquarian Dr Alexander Catcott. As a 'catastrophist' he saw such caves, and the Pleistocene bones they contained, as evidence of the Biblical flood. The cave came to the attention of Cornish miners in the 1790s who were alerted to the presence of 'green veins' thought to contain copper (actually green clay). After fashioning two other entrances, the miners left when no copper was proved, taking with them the best of the stalactites, which may have been procured for the Marquess of Buckingham, the owner of a renowned copper mine at Dodington in the Quantock Hills, who letters show was considering building a grotto in his garden. Others are believed to have reappeared in the caves of Cheddar Gorge. The cave, which had become 'lost' by 1807, was searched-for over many years until it was finally rediscovered by the BEC in 2003. Apparently a nearby resident regularly hears the sounds of horses' hooves on the bridleway echoing beneath his garden. This was the location of the old eastern entrance, which was backfilled with rubbish and dead animals.

Ref: N. Richards & N. Harding, The Rediscovery of Loxton Cavern, BEC Belfry Bull 520 pp 13-19 (2004) & Mendip Underground, MCRA pp 207-208 (2013)

Miners' tally marks in Loxton Cavern. Photo by Nick Richards.

SOUTH CAVITY
Lost Cave
ST 3745 5587 approx. L 40m VR 40m Alt 35m

This 'lost' cave, mentioned in connection with Loxton Cavern, was described by Dr Catcott as *'Some months after this as the same person was digging for ochre about thirty yards south of this cavern; he broke into another Cave (if it be not rather a continuation of this) which I shall distinguish by naming it South Cavity. This cave for forty yards deep in the earth about twenty long, ten broad and not above two high was when he first discovered it nearly full of water.'* The BEC briefly dug a bottle dump at this location before turning their attention to Loxton Cavern. They also recorded a tiny alcove with slickensides nearby which they christened **Crag Hole**, and noted that c.1999 an extension to the cottage just south of the crag uncovered a descending cave passage, which was swiftly filled in.

Ref: A. Catcott, A description of Loxton Cavern Somersetshire together with an account of the Bones and Teeth of an Elephant and some other foreign Animals that were dug out of Loxton Hill (1759) & N. Richards, Personal communication

LOXTON CAVE
Cave
ST 3742 5596 L 157m VR 25m Alt 50m

This interesting cave, discovered by quarrymen in 1862, consists of several small phreatic chambers and grottoes, connected by crawls and squeezes. There are two entrances, both of which are gated. The cave was once well-decorated, and there were even brief plans to open it to the public. However those plans came to nought and sadly few formations now remain intact. Small extensions were made in 1923 and 1946, and in 1956, UBSS discovered a sparse Pleistocene deposit in a boulder chamber located 15m inside the entrance.

Ref: Mendip Underground, MCRA pp 206-207 (2013)

LOXTON SAND QUARRY CAVE 1
Cave
ST 3747 5603 L 22m VR 4m Alt 35m

Located in the south-east corner of a small overgrown quarry adjacent to the Loxton-Christon road, several interconnecting entrances yield a small network of loose, ochre-filled crawls. It is the largest of several small caves excavated by the BEC in 2000-2001 during their search for Loxton Cavern. Some of the connections have since collapsed.

Ref: N. Richards & N. Harding, Loxton Sand Quarry Caves & Loxton Sand Quarry Cave and its environs, BEC Belfry Bull 521 pp 23-26 (2005)

LOXTON SAND QUARRY CAVE 2
Cave
ST 3745 5607 L 3m VR 2m Alt 42m

Located above the quarry in the north-west corner of the plot, this is a small chamber with two entrances.

Ref: N. Richards & N. Harding, Loxton Sand Quarry Caves, BEC Belfry Bull 521 pp 23-26 (2005)

LOXTON SAND QUARRY CAVE 3 — Cave
ST 3747 5605 L 14m VR 3m Alt 45m
Located above the western boundary of the quarry plot, a disused badger hole yielded a small network of tight bedding planes.

Ref: N. Richards & N. Harding, Loxton Sand Quarry Caves, BEC Belfry Bull 521 pp 23-26 (2005)

LOXTON SAND QUARRY CAVE 4 — Cave
ST 3746 5604 L 3m VR 2m Alt 37m
Located in the west face of the quarry adjacent to Loxton Sand Quarry Cave 1, this is probably the remains of a small chamber truncated by quarrying.

Ref: N. Richards & N. Harding, Loxton Sand Quarry Caves, BEC Belfry Bull 521 pp 23-26 (2005)

LOXTON SAND QUARRY CAVE 5 — Cave
ST 3746 5605 L 2m VR 1m Alt 40m
Located 5m above floor-level and reached by an easy chimney in the north-west corner of the quarry, this is a low rectangular passage which quickly becomes too tight to follow. There are several impenetrable tubes in the north face of the quarry.

Ref: N. Richards & N. Harding, Loxton Sand Quarry Caves, BEC Belfry Bull 521 pp 23-26 (2005)

COLDWELL — Spring
ST 3764 5590 Alt 8m
A pond fed by a spring adjacent to the parish church.

LOXTON SETT — Mine
ST 3681 5705
In 1788, four local gentry took out a 21-year lease to mine a recently discovered lead deposit on Loxton Hill. The operation was quickly taken over by the owners of Dodington Mine, who employed Cornish miners captained by Matthew Grose, ostensibly to search for copper. Details are scant but this may be the mine recorded as 37m long in 1788 and entered by a narrow shaft. '*Much lead*' was found along with a little manganese and some ochre, neither of which appears to have been of much interest to the owners. The mine was subsequently taken over by the Foxes of Cornwall but by 1807 their interest appears to have waned when nothing of any significance materialised. Grose, who had been moved sideways following his perceived failure at Garden Mine near Dodington, was probably happy to leave as he wrote '*...ill-natured prejudice in the behaviour of the inhabitants at Loxton and it's vicinity against suffering any but themselves to inhabit the county*'. The NGR is approximate.

Ref: M. Clarke, N. Gregory and A. Gray, Earth Colours – Mendip and Bristol Ochre Mining, MCRA, p 155 (2012)

BANWELL

The caves of Banwell are mainly situated on two distinct hills. Banwell Hill is the long, linear hill to the south of the village, and the Banwell Caves, a 1.7 hectare geological and biological SSSI, are located at the western end. Galena, baryte, manganese and ochre have all been extracted, and the lead shafts were claimed to be '*80 fathoms deep*', which, if true, would place them among the deepest on Mendip. However, the relatively low-key scale of operations in comparison with the more extensive workings on other nearby hills, must cast considerable doubt on this claim. Either way, the mines appear to have been largely unproductive. Apart from the caves the hill is also notable for a number of follies, which at one time included arches, temples, gazebos, obelisks, a sham prehistoric quoit, an ornate pebble summerhouse, an inscribed tombstone (marking the reinternment of a skeleton unearthed by William Beard) and a 20m high tower, known locally as the Banwell Pepperpot! Sadly, not all have survived. The second hill, Banwell Plain, lies to the east of the village and is the site of a deer park constructed by Bishop Beckington (1443-1465). Occupied in the Iron Age, the summit hillfort has yielded flint implements from the Palaeolithic, Neolithic and Bronze Ages. Banwell's other notable surviving feature, is the prominent castle which lies at the top of Banwell High Street. Despite its medieval appearance, it is, in fact, a folly built in 1845, and was once owned by Richard Calvert, the proprietor of the Banwell Ochre Caves. Recently it has been claimed that St Patrick, Ireland's patron saint, was born and bred in the village, which was known to him as Bannaventa. Banwell is probably derived from Bana's Well, an early name for the spring which constitutes Mendip's fourth largest rising.

BANWELL HILL

HILLEND HOLE Cave
ST 3796 5882 L 2m VR 1m Alt 46m

Discovered by engineers constructing the M5 through the western tip of Banwell Hill, this small limestone cavity was located high up on the east side of the motorway cutting. The most northerly of four adjacent holes recorded by the ACG, it was explored and destroyed in 1970.

Ref: C. Richards, A Record of the Caves Discovered During Motorway Construction at Banwell 1970-1971, ACG N/L pp 44-48 (Nov 1972)

HINTON'S HOLE
Cave
ST 3798 5875   L 3m VR 2m Alt 38m
The second of the four holes uncovered during construction of the M5, this small Dolomitic Conglomerate chamber, located low down on the east side of the motorway cutting, was visited by the ACG in 1970, and destroyed shortly after.

Ref: C. Richards, A Record of the Caves Discovered During Motorway Construction at Banwell 1970-1971, ACG N/L pp 44-48 (Nov 1972)

BRIDEWELL CAVE
Cave
ST 3799 5873   L 9m VR 2m Alt 38m
Located low down on the east side of the M5 cutting, this short section of passage, formed in Dolomitic Conglomerate, was intersected at roof level in 1970, and destroyed shortly after.

Ref: C. Richards, A Record of the Caves Discovered During Motorway Construction at Banwell 1970-1971, ACG N/L pp 44-48 (Nov 1972)

WHITLEY HEAD RIFT
Cave
ST 3798 5872   L 8m VR 5m Alt 37m
The last of the four caves discovered during construction of the M5 cutting, this rift, formed in Dolomitic Conglomerate was explored and destroyed in 1971.

Ref: C. Richards, A Record of the Caves Discovered During Motorway Construction at Banwell 1970-1971, ACG N/L pp 44-48 (Nov 1972)

HILLEND BARYTES MINE
Mine
ST 3831 5897
Located just north of 'The Caves' and adjacent to the junction between Well Lane and High Street, a large deposit of barytes appears to have been extracted from fragmentary phreatic hollows or cave passages, which were probably intercepted during quarrying. The workings have all been filled in.

Ref: N. Richards, Personal communication

BANWELL BONE CAVE
Cave
ST 3822 5881   L 100m VR 20m Alt 73m
Also known as Banwell Bone Cavern and The Bone House, this cave is justly famous for its rich deposit of Pleistocene animal bones discovered in the early 19th century, many of which can still be seen in situ. Originally accessed by a (now blocked) entrance located at the western end, the large Bone Chamber is now entered by a set of artificial steps. A short descent gives access to Baker's Extension, while the east end of the chamber is blocked by boulders which prevent a connection with Banwell Stalactite Cave. The cave was accidentally discovered in September 1824 by miners intending to drive a horizontal tunnel into Banwell Stalactite Cave. The earth floor of the cave was found to be littered with animal bones, including bear, reindeer, bison and wolf, which the landowner, George Henry Law, Bishop of Bath and Wells, took to be the remains of

animals drowned in the Biblical flood. A local farmer and antiquarian, William Beard, carried out a major excavation of the bone material in the cave, removing the most interesting specimens and stacking the remainder around Bone Chamber. He went on to act as guide to the cave for over forty years. The cave was opened to the public in 1825, and alongside various follies built by Bishop Law, remained a popular local attraction until 1864. Later excavation by ACG led to the discovery of the Baker Extension in 1952.

Ref: J. Chapman, The Story of Banwell Caves, Banwell Caves Heritage Group (2011) & Mendip Underground, MCRA pp 50-51 (2013)

The Original Bone Stacks, Banwell Bone Cave

BANWELL STALACTITE CAVE
Cave
ST 3830 5879 L 280m VR 75m Alt 79m

Also known as Banwell Stalactite Cavern, this interesting cave consists of a steeply-sloping series of large chambers connected by some particularly unstable boulder ruckles. No active stream is present in the cave, but the lowest chamber contains a deep lake related to the local water table. This quoted grid reference is the commonly used entrance, but there are three other blocked minor entrances located at ST 3829 5881, ST 3826 5883 and ST 3829 5880. Other features of interest include displays of barytes, evident inside phreatic pockets lining the chamber walls, some short, mined passages and the Bishop's Chair, a large fallen boulder with a phreatic half-tube in one face. The cave was opened by miners sometime during the 18th century. For a while it was the talk of the village and one visitor described it as being as *'big as the interior of Banwell Church'*.

Dr Alexander Catcott of Bristol states that it was open in 1756, but later became blocked. The cave was re-entered by miners employed by Rev. Francis Randolph in 1824 and shortly after opened as a show cave. It remained a local attraction until 1864. In 1970, ACG discovered Great Chamber and Green Lake Chamber.

Ref: J. Chapman, The Story of Banwell Caves, Banwell Caves Heritage Group (2011) & Mendip Underground, MCRA pp 50-51 (2013)

Banwell Stalactite Cave. Photo by Steve Sharp

HOLLYBUSH SHAFT

Mine

ST 3863 5865

L 6m VR 6m Alt 91m

Once penetrable to a boulder choke at -6m, this small fenced shaft is now almost full to the brim with tipped rubbish.

Ref: C. Dockrell, Banwell Hill Mines, ACG N/L p 24 (Winter 1987)

TOWER MINE 2
ST 3869 5866
Mine
L 5m VR 5m Alt 100m
This fenced, circular ginged shaft is comprehensively choked.

Ref: C. Dockrell, Banwell Hill Mines, ACG N/L p 24 (Winter 1987)

TOWER MINE
ST 3872 5865
Mine
L 18m VR 15m Alt 95m
A roomy 9m entrance shaft, leads to a small, choked natural chamber, with blocked connections to both Banwell Levvy and Great Maple Mine. The shaft, which is fenced and marked with blue tape, has been partially filled and the entrance to the small natural chamber at the base of the shaft is currently blocked (2018).

Ref: C. Richards, Banwell Levvy and Tower Mine, ACG N/L pp 90-91 (Oct 1968) & C. Dockrell, Banwell Hill Mines, ACG N/L p 24 (Winter 1987)

GREAT MAPLE MINE
ST 3875 5866
Mine
L 71m VR 46m Alt 100m
This interesting example of a Mendip lead mine is believed to date from the 19th century. An 8m deep circular ginged entrance shaft originally yielded a 40m long steeply-sloping tube with a blocked connection with Tower Mine. Small veins of galena were visible, while numerous shot holes, a fragile windlass and the obvious grooves caused by the hauling ropes were among the features of interest. The mine was re-entered by ACG in 1975, but was later permanently closed. Recent investigations (2018) showed that the mine is currently blocked below the entrance shaft, which has now been fenced off and marked with blue tape. The fate of the historic windlass remains unknown.

Ref: Mendip Underground, MCRA p 164 (2013)

BANWELL LEVVY
ST 3875 5863
Mine
L 77m VR 14m Alt 92m
First explored by ACG in 1951, and also known as Sprite's Cave and Howlings Hole, this roomy mine level is presumably the unsuccessful iron trial, driven by '*Mr. Harford of Bristol*' and visited by the Rev. John Skinner in 1833. The tunnel leads to a chamber, with a small extension to the left. Another, very loose, partly-natural tunnel once connected with Tower Mine, but this was collapsed intentionally in 1953 when it was deemed hazardous to local children. Messrs. Harford & Davies obtained a lease for the rights to the minerals on the hill c.1829, following earlier work by Messrs. J. H. Smythe-Pigot and G. Emery.

Ref: M. Clarke, N. Gregory and A. Gray, Earth Colours – Mendip and Bristol Ochre Mining, MCRA, pp 150-152 (2012) & Mendip Underground, MCRA p 51 (2013)

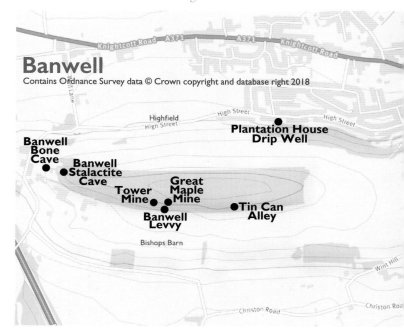

TIN CAN ALLEY   Mine
ST 3898 5863   L 26m VR 12m Alt 95m

Located close to the edge of the wood, a 10m deep ginged shaft descends to a small chamber containing another capped shaft to the surface and a blocked continuation below. It is currently covered with metal and surrounded by a fence festooned with blue and white tape.

Ref: C. Dockrell, Banwell Hill Mines, ACG N/L p 24 (Winter 1987)

BANWELL COLLAPSE   Mine
ST 3925 5885   L 7m VR 3m Alt 80m

A small collapse, which yielded 4m of horizontal passage. Possibly a trial iron working, it was recorded by ACG in 1987 and subsequently capped.

Ref: Banwell Collapse, ACG N/L (Spring 1987)

PLANTATION HOUSE DRIP WELL   Cave
ST 3915 5898   L 35m VR 7m Alt 46m

This unusual cave is situated in the back garden of Plantation House. Behind a locked door, a flight of steps enters a roomy, L-shaped natural passage, which has been much-modified, initially by mining, and then during construction of a large brick-lined tank, which remains in situ. This structure was built to provide clean

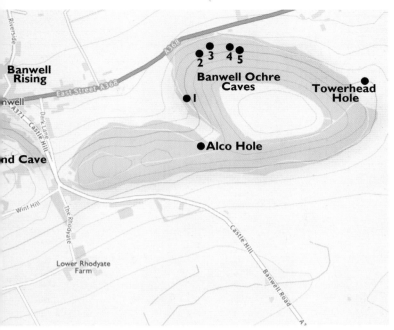

cooking and drinking water to the house, while additional water for cleaning and bathing was collected from the roof. There is also a rusty pump, which is apparently still used to water the garden. The cave was investigated by John Tucker in 1984, who noted some modern animal bones above a thick stalagmite floor.

Ref: J. H. Tucker, Plantation House Drip Well, Banwell Soc Arch Jnl (Search) 20 pp 4-11 (1984), Plantation House Drip Well – Final Report, Banwell Soc Arch Jnl (Search) 21 pp 54-55 (1986) & K. A. B. Edwards, The Cave at Plantation House – A Domestic Appendix Banwell Soc Arch Jnl (Search) 20 p 12 (1984)

BANWELL HIGH STREET MINE — Mine
ST 3934 5899 L 6m VR 1m Alt 40m

This small iron mine was uncovered in 1952 by workmen digging the foundations for the sixth in a row of new council houses. It was investigated by ACG who followed a small passage into an ochreous chamber containing pick marks. The council houses were built on the site of some old miners' cottages and a collapsed continuation led towards a small limestone cliff, where a choked entrance had formerly been visible. The workings were quickly filled in. The date of this working is unknown although a George Webb apparently '*did a good business*' in this vicinity during the 1850s.

Ref: J. W. Hunt, Mine at High Street, Banwell, ACG N/L p 118 (Dec 1968) & M. Clarke, N. Gregory and A. Gray, Earth Colours – Mendip and Bristol Ochre Mining, MCRA, p 152 (2012)

BANWELL RISING
ST 3988 5915

Rising
L 11m VR 11m Alt 10m

Derived from Bana's Well and also known as Banwell Spring, the rising that gives the village its name, is the fourth largest in the Mendip Hills and has never been known to fail. It gives approximately 2.3 million gallons a day, rising to 7 million two days after heavy rain. The spring originally surfaced in an attractive pool of blue-green water from an 11m deep rubble-filled pot. Sadly, the pool is long gone, and the water is now contained within a massive concrete structure which was covered over by the village bowling green when the spring was capped in 1931. It appears to derive from two distinctive sub-catchments, which lie to the east and west of the spring and drains almost the entire northern limb of the Blackdown pericline - a catchment area estimated at 14 square kilometres. Proven feeders include Rowberrow Swallet (7 days) and Swan Inn Swallet (50-150 hours), some 5.5 km distant, along with the underground streamway in Mangle Hole (37 hours). Small pieces of coal are reputedly thrown up by the spring and legend has it that the Shipham dowsers claimed to have traced the course of the Banwell Spring back to a spot in Longbottom Valley near Shipham, where, in 1813, a shaft was sunk in search of coal. However, attempts to prove a connection with the area (including the stream seen in Singing River Mine) have so far failed, and the 'coal' is probably the residue of clinker used during construction of the pond. Historically the spring has been put to many uses. Collinson, writing in 1791, says that the spring was medicinal, and that it was once famous for curing scrofulous disorders, while there have been suggestions that the

Banwell Pond c.1905

large pool it once fed was used as the local 'Ducking or Cucking Stool', a punishment meted out to nagging women and dishonest tradesmen. The water - which once supplied a brewery, an extensive writing paper factory, a lemonade factory and a flour mill - is so pure that it has even been used in the manufacture of Bank of England notes. It is the source of the River Wick, or Banwell River, which discharges into the Bristol Channel near Woodspring Priory. Another spring enters the river through a pipe beside the pump house and there are said to be other springs in gardens nearby. **Adam's Well** (ST 398 590), was once esteemed for its '*efficacy in scrofulous disorders*' but is now lost beneath cottages.

Ref: S. L. Hobbs & P. L. Smart, Delineation of the Banwell Spring Catchment Area and the nature of the spring hydrograph, UBSS Proc 18 (3) pp 359-366 (1989) & E. K. Tratman, Hydrology of the Burrington Area, Somerset, UBSS Proc 10 (1) pp 22-57 (1962-63)

SAND CAVE
Cave
ST 3986 5886   L 4m VR 4m Alt 65m

Also known as Footprint Cave and Sandy Hole, this cave was probably blasted open by quarrying c.1890-1900. Located 10m below the lip of the south side of the main quarry, the cave consisted of a small chamber with an impassable chimney leading to the surface. According to local folklore, on entry, the clear impression of a human footprint could be seen on the sandy floor. On the northern side of the quarry, ACG noted several other small caves, but all were permanently lost when the entire quarry was later infilled with rubbish.

Ref: J.W. Hunt, The Sand Cave, Banwell Quarry, ACG N/L pp 6-8 (Dec 1990)

WINTHILL SHAFT
Mine
ST 3974 5845 approx   L 12m VR 12m Alt 55m

The area around Winthill was extensively mined and shafts have suddenly opened up on more than one occasion. This is also the site of an important Romano-British settlement and archaeological investigations carried out by ACG & AS during the 1950s uncovered a vertical shaft sunk into the solid limestone rock. It was covered over with large rocks but contained nothing beyond clay, pieces of slate and portions of iron bar which were found lodged in the side. Several old skeletons lay adjacent to the shaft, which appears to have been sunk right through the middle of a burial ground. Some of the bones are believed to have been reinterred inside the shaft when it was filled in although others were removed to Axbridge Museum. Radiocarbon dating produced results ranging from AD 430-810. The most important find however was the famous 'Winthill Bowl', a 4th century glass drinking vessel inscribed in Latin with the term '*Long life to you; drink, and good health*'. It is now in Oxford's Ashmolian Museum. There are also several wells in Winthill which are said to have been sunk by the Churchill ochre miners, Richard and Owen Heal.

Ref: J.W. Hunt, Report on the Society's 1956 Excavations at the Winthill Site, Banwell, ACG & AS Jnl Vol 3 No 2, pp 5-9 (1957)

YARBOROUGH FISSURE
Cave
ST 3883 5817
L 3m VR 3m Alt 29m

Located below a small cliff opposite a T-Junction, and covered by a pallet and metal farm gate, this small vertical fissure, in Dolomitic Conglomerate, is now largely filled with leaf mould. Also known as Yarborough Rift and Elborough Rift, it was apparently dug archaeologically in 1951 but nothing is known of the few bones which were said to have been recovered. In a field known as Stone Hams, located some 400m to the south-east, is a large prehistoric standing block of conglomerate, some 1.5m wide x 2.5m high, which is said to have been hurled here from the top of Crook Peak during a competition between the Devil and a local strongman. There were once many standing stones on Mendip, which were often used to denote parish boundaries, but sadly most, including the evocatively named Sliperstone, Petra Perforata, Hob-in-the-Moss and The Long Man, were removed during wartime drives to increase productivity.

Ref: N. Barrington, W. Stanton, The Complete Caves and a view of the hills, p 182 (1977)

BANWELL PLAIN

Among the many small pits and workings spread out across the northern flank of Banwell Plain lie five significantly larger sites, collectively known as the Banwell Ochre Caves. There was apparently a further site, which was regarded as too dangerous by the miners, who subsequently filled it in. These part-mine, part-natural phreatic tunnels, were originally filled with ochreous earth, extraction of which was in progress right up until recent times. The caves contain the most extensive and accessible yellow ochre workings in the Mendip area and a wide variety of ochre types and iron hydroxides (limonites) can be examined in situ. They appear to be associated with the unconformity at the base of the Dolomitic Conglomerate and the evidence of their accumulation as residual ore-bodies associated with ice age (Pleistocene) sediments is so well displayed that the caves have been assigned SSSI status. Although ochre has been mined in this general area for several centuries, the workings described here were excavated by miners working for the Mendip Oxide & Ochre Co. beginning in 1934. The mines yielded over 5000 tons of ochre which was moved by a narrow-gauge railway and a tramway to the road for transportation to the Golden Valley works at Wick. Mining finally ceased c.1952. Late Pleistocene bones have been reported and a 'much mineralized tusk' was presented to Wells Museum in 1938.

Banwell Plain from Sandford Hill with Banwell Hill, Brean Down and Steep Holm beyond. Photo by Rob Taviner

BANWELL OCHRE CAVE 1
Mine
ST 4060 5904   L 62m VR 12m Alt 43m

Also known as Pool Mine, this site lies close to the track on the west edge of the wood. A low entrance passage leads into roomy upper and lower galleries. This mine was once used to store the empty mining drams and there are rails and stemples still in situ.

Ref: Mendip Underground, MCRA p 52 (2013)

BANWELL OCHRE PIT
Mine
ST 4062 5919   L 7m VR 5m Alt 45m

A blind vertical ochre pit in undergrowth between the main forest track and Banwell Ochre Cave 1.

BANWELL OCHRE CAVE 2
Mine
ST 4066 5923   L 154m VR 25m Alt 53m

Also recorded as Big Cave, The Ochre Cave, Horse Trail Mine, Mud Slide Mine and Banwell Ochre Cave, this cave, discovered in 1937, consists of two large parallel passages accessible from three entrances. Fossilised bird's nests of mud, probably of martins, were found on the ledges surrounding the Upper Entrance. Immediately above the Upper Entrance lies the High Series (also recorded as Lesser Ochre Cave), an apparently separate working, where a roomy tunnel can be followed for over 20m to a choke. A short passage on the right near the entrance yields a letterbox squeeze into a roomy chamber with an impassable connection to a balcony in the main cave below.

Ref: Mendip Underground, MCRA p 52 (2013)

Banwell Ochre Cave 1937. Dave Irwin Collection

BANWELL OCHRE CAVE 3
Mine
ST 4067 5927 L 92m VR 15m Alt 40m

Also known as the Middle Series and located approximately 40m north of Banwell Ochre Cave 2, this cave also consists of upper and lower galleries, connected by climbs and pitches. The ochreous earth in this cave is locally overlain by up to 2m of fine-grained, wind-blown quartz sand.

Ref: Mendip Underground, MCRA pp 52-53 (2013)

BANWELL OCHRE CAVE 4
Mine
ST 4073 5926 L 62m VR 10m Alt 46m

Also known as Main Entrance, a roomy entrance cutting leads to a series of irregular strike galleries excavated on one level.

Ref: Mendip Underground, MCRA p 53 (2013)

BANWELL OCHRE CAVE 5
Mine
ST 4075 5925 L 31m VR 7m Alt 52m

Located roughly 20m east of Banwell Ochre Cave 4, the working consists of a simple gallery, with a step-up to the left into a passage which emerges at a second entrance immediately adjacent to the first. The main entrance has recently been affected by a collapse.

Ref: Mendip Underground, MCRA p 53 (2013)

ALCO HOLE
Mine
ST 4066 5883 L 7m VR 4m Alt 70m

Recorded by the BEC in 2007, this is a narrow mined ochre rift choked with rocks and gin bottles at a depth of 4m. The ruins of the house nearby may have been built using stone taken from a substantial Elizabethan dwelling owned by Godwyn, Bishop of Bath and Wells from 1584 to 1590.

Ref: N. Richards, Personal communication

TOWERHEAD HOLE
Cave
ST 4133 5914 L 3m VR 1m Alt 47m

Located by the side of the sunken track immediately north of a small quarry, a low, bramble-mired arch enters a tiny circular chamber.

Ref: N. Barrington, W. Stanton, The Complete Caves and a view of the hills, p 163 (1977)

SANDFORD

The summit of Sandford Hill is a broad plateau pitted with long rakes and deep lead mines. The mines were claimed to have been worked to a depth of up to 80 fathoms (480 ft), below which the '*water flows in so fast that it cannot be pumped*'. Over the course of centuries, several natural caves were encountered, including two which contained notable bone deposits, and a third known as the Sandford Gulf, a legendary lost shaft described as being '*240 feet deep*'. Yet more natural cavities were found on the wooded slopes surrounding the summit when these were later worked for ochre. Several surveys of the hill have been carried out over the years, including the Swarbrick brothers of Sidcot School (1948-51), Mendip Nature Research Committee (1955), Sidcot School Speleological Survey (1959), Shepton Mallet Caving Club (1990), Axbridge Caving Group (2004) and subsequent follow up work conducted by the MCRA in 2014-2016. These sites are referenced with the notations SWARBRICK, MNRC, SSSS, SMCC, ACG and MCRA respectively. There has been a good deal of duplication, and some sites - particularly the myriad of small mines on the northern slopes of the hill - remain open to interpretation. Early mining records are scarce, although it is known that William Jefferies, acting on behalf of Dr Alexander Catcott, visited this hill in 1770 to procure some animal bones which had been discovered buried in a fissure. He managed to extract a few specimens, but work was curtailed due to the animosity of the miners. In 1829, Rev. David Williams brought 'The Gulf' to our attention, followed a year later by the Rev. John Skinner, who recorded that work was underway in Sandford Levvy. William Beard excavated the Sandford Bone Fissure in 1838. By 1861 all mining had ceased, but the outbreak of World War I created a surge in demand for native ochre, which was used in the production of camouflage colours. This prompted the final development on the hill, which consisted of a series of small ochre pits with relatively shallow underground workings, exemplified by the numerous sites still visible today on the lower northern slopes. Small-scale mining continued into the 1940s and the last major mine on the hill (Sandford East Ochre Mine), ceased working shortly after WWII. Lead and ochre weren't the only important riches unearthed by the miners, for around 1830, a man appropriately named Rich, '*drove his pick into pot of coins. Further search revealed three crocks*'. The hoard, numbering almost 300 denarii, dating mainly from the 3rd century A.D. was discovered on the north side of the hill close to Sandford Levvy and was mostly distributed among family members. Almost the entire hill was once under significant threat of being quarried away, but thankfully, following a deal brokered in the 1990s to extend Whatley Quarry, the owners relinquished their rights to Sandford Quarry, saving the remaining part of the hill, including all the known caves and mines, from an ignominious fate. Many sites of interest remain, and these are broadly aligned along three parallel paths, which lead west along the hill from the ski centre.

LOWER PATH

From the ski centre car park, follow the road which leads uphill to the west towards Lyncombe Lodge. At the first hairpin, take the obvious public footpath on the right alongside an outbuilding and keep to the lower path (marked with a yellow disc). This contours the northern flank of the hill close to the edge of the wood. With the exception of King Mine, the most important sites lie to the south (uphill) of the path.

KING MINE
Mine
ST 4301 5937   L 120m VR 25m Alt 55m

Marked on the SMCC Survey as SMCC-3, this small mine working, entered by a 7.5m shaft at the foot of Sandford Hill, intercepts a significant natural chamber 6m high, 7m wide and 15m long. It contains evidence of the mining techniques used and some good dog-tooth crystals. There also appears to be a significant newt colony in the vicinity of the entrance. The mine was rediscovered in 2005 by Dave Upperton of WCC who named it in honour of his friend, Lt Tony King, of 849 Naval Air Squadron, who was killed during the 2003 Iraq War.

Ref: Mendip Underground, MCRA p 188 (2013)

PSEUDO UDDERS HOLE
Mine
ST 4299 5933   L 12m VR 4m Alt 66m

Marked on the SMCC Survey as SMCC-2. Two steeply-descending rifts connect at the bottom to form a short through trip.

Ref: E. Waters, Of Caves, Mines and Legendary Gulfs, SMCC Jnl 12 (1) pp 8-16 (2007)

MCRA-SH-1
Mine
ST 4296 5934   L 4m VR 3m Alt 66m

A small shaft yields a choked sloping rift.

SWARBRICK-A-8
Lost Mine

Described in the Swarbrick survey as a *'model of the shallow type of mine'*, all that is known about this site is that it was located to the east of Sandford Levvy at more or less the same level. Sadly, it can no longer clearly be recognised, but when explored c.1948, a loose narrow slit was descended for 4m to a platform from where a further narrow vertical squeeze and 2m drop gained a small chamber. Fine dog-tooth crystals were in evidence and a small passage continued for a short distance to a roof fall.

Ref: T. G. & J.T. Swarbrick, A Survey of the Lead, Ochre and Calamine Mines on Sandford Hill, pp 33-34 (1948-1951)

SWARBRICK-A-7
Lost Mine

Another lost site located to the east of Sandford Levvy, this mine was described as a narrow, unstable level that had been driven into the hill. Its exact position is unknown.

Ref: T. G. & J.T. Swarbrick, A Survey of the Lead, Ochre and Calamine Mines on Sandford Hill, p 31 (1948-1951)

SMCC-4

Mine

ST 4292 5936   L 24m VR 4m Alt 54m

Also logged as ACG-01 and probably Swarbrick-A-6, this square shaft lies about 20m east of Sandford Levvy. From the bottom a tight rift leads past crystals to connect with Sandford Levvy close to the entrance. This was the entrance used by a Home Guard Auxiliary Unit during WWII after they had blocked the main adit. Both ends of the passage can become blocked.

Ref: E. Waters, Of Caves, Mines and Legendary Gulfs, SMCC Jnl 12 (1) pp 8-16 (2007)

SANDFORD LEVVY

Mine

ST 4291 5936   L 588m VR 43m Alt 55m

Marked on the SMCC Survey as SMCC-5 and also known as The Levvy, Sandford Hill Lead Adit, ACG-02 and Swarbrick-A-5, the main feature of this mine, which began life c.1830 under the supervision of Captain William Webster, is a horizontal mined passage, 313m long, which intersects two important crossroads. The first offers two interesting detours, especially the left-hand branch, where some difficult climbing enters extensive 18th century workings which lie close to and are doubtless associated with Pearl Mine and the Groof House and Fern series of mineshafts (see below) situated on the hillside above. A connection here offers the potential for a fascinating through trip. Climbing in the vicinity of the Second Crossroads has yielded more passage and the impressive 'Great Rift has been cited as a possible contender for the legendary lost 'Sandford Gulf' described in Netherworld of Mendip (1907) as follows. '*In driving an extensive level through a hill, 60 fathoms below the summit the miners came across a gigantic rift. A man was let down on a long rope but was unable to see walls or sides of this tremendous abyss.*' Stanton, and others, argued that this could only refer to the Great Rift, the theory being that the fissure once continued down well below the current floor level and had been subsequently infilled with the rubble extracted from the continuation of the adit (estimated by Stanton at 1000 tons). The biggest problem with this theory was that 'The Gulf' was first mentioned very early in 1829 - the same year that Webster was granted official permission to commence work on Sandford Hill – and it seemed impossible that tunnelling could have reached as far as the Second Crossroads. However it was not uncommon for mining leases to be granted in retrospect, so it is eminently possible that work was well underway prior to the 'official' recorded date. A find of galena was reported and small veins of zinc blende can still be found, but no record of commercial success exists. The mine was later used by an Auxiliary Unit of the local Home Guard during the Second World War who sealed up the main entrance, choosing instead to enter via a gruff located '*some way off*'. This was probably SMCC-4.

Ref: Mendip Underground, MCRA pp 306-309 (2013) & R. M. Taviner, The Sandford Gulf and other lost caves of Sandford Hill, (2016), available online via www.mcra.org.uk

- Hutton Cavern - Catcott (1757)
- Beard's Second Cavern (1833)

Bleadon Cavern

x - 1757 entrance
E - Modern entrance

BEARD'S SECOND CAVERN OF BONES (1833)   Cave

In 1832, Beard once again began excavating shafts on Hutton Hill, and in 1833, yet another bone cavern was discovered. Beard employed John Heal of Shipham to produce an accurate dialling survey of the cave, which can still be viewed today in the Somerset Heritage Centre. The survey notes prove that this cave was originally entered from Hutton Parish, but to help facilitate the removal of the bones, at least one other shaft was subsequently sunk over the wall in Bleadon parish. The bones are housed in the Museum of Somerset, and have been dated to 250,000 BP. It's now known that this 'third' cave was in fact Catcott's original Hutton Cavern, the one known to us today as Bleadon Cavern.

The 1970s

During the 1970s, ACG and Hutton Scouts began searching for the famous 'Lost Cave of Hutton' and uncovered several caves which appeared to at least partly tally with the antediluvian accounts. Bleadon Cavern was opened in 1970, and was quickly heralded as the famous lost cavern, but the subsequent failure to find anything resembling the three vertical chambers described by Williams (then considered part of the same system), eventually forced the discoverers to change their opinion, deciding that the new cave must instead be the lesser of the two caves described by Catcott.

Digging continued, and several further caves were uncovered, including Primrose Cave and May Tree Cave (which they recorded as Hutton Cavern 3), and Well Shaft Cave (recorded as Hutton Cavern 4). However, the continuing failure to find anything satisfactorily resembling the lost cave described by Rutter inevitably caused interest to wane, and operations finally drew to a close in 1977.

PRIMROSE CAVE (ST 3602 5815) – see UPPER CANADA CAVE

MAY TREE CAVE (ST 3603 5816) – see UPPER CANADA CAVE

The 2000s

Beginning in 2006, ACG once again showed an interest in the area and several more caves were discovered, including, crucially, Upper Canada Cave. Subsequent work with a mechanical digger, uncovered a broad pit surrounded by several entrances, three of which appeared to correspond so closely with the Rutter elevation, that the team confidently felt that they had finally found the fabled lost cavern. However, a presentation to a packed house at Wells Museum failed to convince the jury, and further doubts were raised when almost all the newly extracted bones were shown to be from relatively modern carcasses. However, subsequent analysis of the various accounts led this author to suggest that the descriptions provided by Catcott and Williams were so different that they could only be two completely different caves. With this realisation, the rest of the pieces quickly fell into place and it is now clear that Bleadon Cavern (Beard's Second Cavern of Bones) is Catcott's original lost cavern, while Williams rediscovered Catcott's lesser cavern (Hutton Cavern 2), which now forms part of Upper Canada Cave. The famous Williams elevation produced in Rutter is in fact a representation of the May Tree Cave (B) and Primrose Cave (C) entrances into the system, along with Archway Cave (A). It should be recorded that practically all the important pieces of the 'lost cave' were discovered during the 1970s, with the later work providing the crucial evidence needed to finally help place them together in the correct order.

BLEADON CAVERN
Cave
ST 3607 5814
L 335m VR 44m Alt 111m

Recorded at various times as Hutton Cavern, Bleadon Cave, Hutton Cave and Beard's Second Cavern of Bones (and also confusingly as Hutton Cavern II and Maytree Cavern!), this is definitely the cave discovered by William Glisson and his ochre miners between 1739 and 1746 (the latter being the date of a smoke mark). Passing directly beneath the parish boundary wall, it was the largest of several natural caves and fissures opened during mining operations in the area. Subsequently lost, it was re-entered by Beard in 1833,

Hancock's Shaft, Sandford Levvy. Photo by Ed Waters

MCRA-SH-8
ST 4291 5933
Mine
L 4m VR 4m Alt 65m

A vertical shaft in a narrow rift.

MCRA-SH-7
ST 4289 5934
Mine
L 2m VR 1m Alt 65m

A low bedding with a notable red ochre-stained roof.

SMCC-7
ST 4287 5935
Mine
L 7m VR 2m Alt 55m

Two, interconnecting small holes, located a few metres west of Sandford Levvy and also logged as ACG-03, ACG-04 & Swarbrick-A-4. A small depression partway along the connecting passage may be the top of a blocked shaft.

Ref: E.Waters, Of Caves, Mines and Legendary Gulfs, SMCC Jnl 12 (1) pp 8-16 (2007)

MCRA-SH-2
ST 4287 5934
Mine
L 2m VR 1m Alt 57m

A tiny sloping tube.

ACG-05   Mine
ST 4286 5936   L 2m VR 2m Alt 54m
Also recorded as Swarbrick-A-3, this is a small vertical pit cut through solid rock.
Ref: ACG Logbook No.9 pp 15-22 (2004)

MCRA-SH-3   Mine
ST 4286 5934   L 3m VR 3m Alt 57m
A small, ginged shaft leads to a choked rift. Also logged as ACG-06.
Ref: ACG Logbook No.9 pp 15-22 (2004)

MCRA-SH-4   Mine
ST 4286 5932   L 2m VR 2m Alt 64m
A small choked vertical pit.
Ref: ACG Logbook No.9 pp 15-22 (2004)

SMCC-8   Mine
ST 4282 5935   L 2m VR 2m Alt 52m
A choked, 2m deep ginged shaft. Also logged as ACG-08.
Ref: E. Waters, Of Caves, Mines and Legendary Gulfs, SMCC Jnl 12 (1) pp 8-16 (2007)

ACG-07   Mine
ST 4283 5936   L 2m VR 2m Alt 57m
Another blocked shaft, a mere 2m deep

Ref: ACG Logbook No.9 pp 15-22 (2004)

SMCC-10   Mine
ST 4280 5932   L 10m VR 4m Alt 59m
A sloping rift leads to a choke below an impassable skylight.
Ref: E. Waters, Of Caves, Mines and Legendary Gulfs, SMCC Jnl 12 (1) pp 8-16 (2007)

SWARBRICK-B-4   Lost Mine
This lost site consisted of a roomy triangular chamber (6m long x 3m wide x 2m high) with two entrances in the eastern and western corners and a further very narrow entrance to the north.
Ref: T. G. & J. T. Swarbrick, A Survey of the Lead, Ochre and Calamine Mines on Sandford Hill, p 37 (1948-1951)

BAT HOLE   Cave
ST 4277 5934   L 15m VR 3m Alt 57m
Marked on the SMCC Survey as SMCC-11 and also recorded as SR Shaft, ACG-09 and Swarbrick-B-5, this is a large hole just south of the main path at the bottom of the hill. A steep debris slope leads into a natural chamber modified by miners. When explored c.1948 the entrance cutting was described as '*10 feet long x 8 feet high*', which led into a chamber

'about 6 feet high'. It has since been filled in considerably.

Ref: E. Waters, Of Caves, Mines and Legendary Gulfs, SMCC Jnl 12 (1) pp 8-16 (2007), T. G. & J. T. Swarbrick, A Survey of the Lead, Ochre and Calamine Mines on Sandford Hill, pp 40-41 (1948-1951) & S. E. Cowdry & R. H. Hotston, 1990-1991 Review of Sandford Hill Mineral Workings, Russell Society Newsletter pp 12-18 (1991)

SMCC-12 — Mine
ST 4277 5932   L 2m VR 2m Alt 59m

A small square shaft, partially hidden beneath a sheet of tin (2014). Also logged as Well Shaft and ACG-10.

Ref: E. Waters, Of Caves, Mines and Legendary Gulfs, SMCC Jnl 12 (1) pp 8-16 (2007) & S. E. Cowdry & R. H. Hotston, 1990-1991 Review of Sandford Hill Mineral Workings, Russell Society Newsletter pp 12-18 (1991)

MANGLE HOLE — Cave
ST 4271 5932   L 350m VR 53m Alt 60m

Logged as SMCC-13 and ACG-11, a deep entrance pitch provides a rather technical descent to an impressive chamber - roomy enough for some commentators to suggest that it should be considered as another (albeit unlikely) candidate for the lost 'Gulf'. Below, several routes unite at Aldermaston Chamber, which contains two deep pools and an underwater streamway which Stanton tested to Banwell Spring in 1977 (37 hours). Although it lies in an area of intense mining activity, the cave is entirely natural (but must have been known to the miners). The cave is very muddy throughout and there are indications that the lowest levels of the cave may flood to a height of 7m. There are many jammed boulders which demand great respect. The cave was discovered and explored as far as Main Chamber by two London cavers in 1970. SSSS continued exploration during the same year and the final Aldermaston Chamber was found by AMCC in 1973. CDG explored the upstream and downstream sumps in 1977 and extended the downstream sump to its present limit in 1987.

Ref: Mendip Underground, MCRA p 211-215 (2013)

SMCC-14 — Mine
ST 4268 5929   L 10m VR 8m Alt 70m

An 8m deep shaft reaches a low earthy chamber. The east wall of the shaft is ginged, the rest is solid.

Ref: E. Waters, Of Caves, Mines and Legendary Gulfs, SMCC Jnl 12 (1) pp 8-16 (2007)

HOLLY SHAFTS — Mine
ST 4267 5931   L 12m VR 5m Alt 60m

Marked on the SMCC Survey as SMCC-16, these are two roomy 5m deep shafts located 40m west of Mangle Hole. A reported connection is currently choked (2016). Also logged as ACG-12 and ACG-13, one of the shafts may be Swarbrick-B-6. ACG-14 is probably a large depression 8m further east.

Ref: E. Waters, Of Caves, Mines and Legendary Gulfs, SMCC Jnl 12 (1) pp 8-16 (2007)

SMCC-17
Mine
ST 4265 5931
L 22m VR 8m Alt 59m

Also logged as ACG-17, an 8m deep shaft leads to 15m of mined passage. Pick marks and the skeletons of small animals are visible. Depressions nearby have been logged as ACG-15, ACG-16 and ACG-20. ACG-15 was described as 5m deep with 3m of small rift below. The rift is now choked.

Ref: E. Waters, Of Caves, Mines and Legendary Gulfs, SMCC Jnl 12 (1) pp 8-16 (2007)

SMCC-15
Mine
ST 4264 5927
L 10m VR 6m Alt 90m

A 4m deep shaft leads to a short passage heading west. There is a short climb down to a lower passage with a stone wall. This is probably the site the Swarbrick brothers recorded as Rift B No.8, which has a similar description but appears to have been slightly longer in their day (1948-51).

Ref: E. Waters, Of Caves, Mines and Legendary Gulfs, SMCC Jnl 12 (1) pp 8-16 (2007), T. G. & J. T. Swarbrick, A Survey of the Lead, Ochre and Calamine Mines on Sandford Hill, pp 46-47 (1948-1951)

SMCC-18
Mine
ST 4263 5930
L 10m VR 3m Alt 75m

Two 3m deep holes lead into a steeply-sloping rift. The rift continues unroofed to the west for several metres. Also logged as ACG-18, ACG-19 and ACG-21 and probably Swarbrick-B-7.

Ref: E. Waters, Of Caves, Mines and Legendary Gulfs, SMCC Jnl 12 (1) pp 8-16 (2007), T. G. & J. T. Swarbrick, A Survey of the Lead, Ochre and Calamine Mines on Sandford Hill, pp 44-45 (1948-1951)

MCRA-SH-5
Mine
ST 4261 5929
L 7m VR 3m Alt 75m

A narrow rift leads to a choke. There are faint initials inscribed in carbide on one wall. This may be the site logged as both ACG-23 and Swarbrick-B-9.

MCRA-SH-6
Mine
ST 4261 5929
L 2m VR 2m Alt 75m

A narrow rift, this is possibly the site logged as ACG-22.

SMCC-19
Mine
ST 4257 5929
L 3m VR 3m Alt 75m

Two 3m deep rifts located a few metres apart in the same vein. They do not connect. This is probably the same site logged as ACG-24.

Ref: E. Waters, Of Caves, Mines and Legendary Gulfs, SMCC Jnl 12 (1) pp 8-16 (2007)

ACG-26
Mine
ST 4257 5933
L 6m VR 6m Alt 55m

Located 3m north of the path, in an area of gruffy ground also known as SMCC-20, this is the deepest accessible point in a rake, 4m long and 1m wide.

Ref: ACG Logbook No.9 pp 15-22 (2004)

ACG-27
Mine
ST 4259 5932 L 6m VR 4m Alt 55m
Located close to the path and due east of ACG-26, in an area of gruffy ground. Also known as SMCC-20, a triangular entrance descends steeply to a tight squeeze and narrow rift.
Ref: ACG Logbook No.9 pp 15-22 (2004)

AUGER'S MINES
Mine
ST 4239 5929 L 4m VR 1m Alt 85m
Named after Richard Auger, this is a collective name for a group of small pits and opencast workings located low down on the north-west slope of the hill. Among the final ochre deposits to be exploited, at least one of the workings remains penetrable - a low crawl which quickly terminates in a small ochreous chamber. This is probably the site that Peter Burr christened Cherry Tree Mine.
Ref: P. Burr, Mines and Minerals of the Mendip Hills, Volume I, MCRA p 598 (2015)

CHAPEL RIFT
Mine
ST 4202 5926 L 6m VR 6m Alt 40m
Also known as Chapel Cottage Rift, this is a small mine working, partly in limestone and partly in Dolomitic Conglomerate, located 10m from the front door of a large house (Chapel Cottage) next door to the Methodist chapel. The south end of the rift is blocked with huge boulders.
Ref: ACG Logbook No.7 p 16 & p 24 (1990)

CENTRAL PATH

Start along the lower path then take the obvious path to the south which leads steeply up the hillside into the wood. Do not cross the stile into the field, but instead continue west through the woods, passing several mineshafts, until an old wall is reached. Cross over to reach a broad path running through the centre of the woods. Many of the sites are located along the northern edge of this path. Alternatively, follow the track which leads south from the ski centre and then west along the open summit of the hill and then turn right as soon as the track re-enters the wood. The path turns sharp left towards Pearl Mine, passing besides several choked mineshafts.

SSSS-43-G
Mine
ST 4314 5927 L 7m VR 7m Alt 75m
Located immediately adjacent to the path this is a narrow shaft, partially choked with boulders and earth.
Ref: SSSS, Sandford Survey Notes (1959-63)

MCRA-SH-14
ST 4325 5924
A tiny choked hole in a rake.

Mine
L 1m VR 1m Alt 100m

MCRA-SH-13
ST 4323 5924
A large fenced-off mineshaft choked at a depth of 5m.

Mine
L 5m VR 5m Alt 100m

ELLIS' WONDER
ST 4315 5922

Mine
L 13m VR 13m Alt 105m

Marked on the Sidcot School Sandford Hill Speleological Survey of 1959 as SSSS-43-M and once 13m deep, this shaft is now just 7m deep (2014). It is believed to be blocked by a large boulder that was once jammed across the middle of the shaft a short distance below the entrance.

Ref: SSSS, Sandford Survey Notes (1959-63)

GRAN'S DISCOVERY
ST 4320 5922

Mine
L 7m VR 7m Alt 105m

Marked on the Sidcot School Sandford Hill Speleological Survey of 1959 as SSSS-43-L, and once 7m deep, this roomy, holly-covered shaft is currently choked with debris at -4m (2014).

Ref: SSSS, Sandford Survey Notes (1959-63)

SAVILLE ROW SHAFT 3
ST 4305 5920

Mine
L 10m VR 9m Alt 113m

Just over 200m east of Pearl Mine, and probably on the same mineral vein, lie four adjacent shafts collectively known as the Saville Row Shafts. First described by SSSS in 1950 and recorded as MNRC-1 to MNRC-4, they are listed on the SMCC Survey as SMCC-1 and on the Sidcot School Sandford Hill Speleological Survey of 1959 as SSSS-43-H, I, J and K. Saville Row Shaft 3 (MNRC-3, SSSS-43-K or New Shaft) connects to Whispering Gallery (Saville Row Shaft 2, MNRC-2 or SSSS-43-J) above the steeply-sloping rubble-floored passage.

Ref: WNHAS & MNRC Report 46/47 pp 29-33 (1955/1956), SSSS, Sandford Survey Notes (1959-63) & E. Waters, Of Caves, Mines and Legendary Gulfs, SMCC Jnl 12 (1) pp 8-16 (2007)

WHISPERING GALLERY
ST 4303 5918

Mine
L 37m VR 31m Alt 113m

Also known as Saville Row Shaft 2, MNRC-2 and SSSS-43-J and first explored by SSSS, a 9m deep shaft reaches a steeply-sloping rubble-floored passage which descends to a depth of 31m. There is an aural connection at the bottom with a side passage at the bottom of Shaft 120. This may also be Swarbrick-C-5, a shaft described as 25m deep with no side passages which ended in a small chamber with a steeply-sloping debris floor.

Ref: T. G. & J. T. Swarbrick, A Survey of the Lead, Ochre and Calamine Mines on Sandford Hill, p 61 (1948-1951), SSSS, Sandford Survey Notes (1959-63), WNHAS & MNRC Report 46/47 pp 29-33 (1955/1956) & E. Waters, Of Caves, Mines and Legendary Gulfs, SMCC Jnl 12 (1) pp 8-16 (2007)

Whispering Gallery and Shaft 1, Saville Row. Photo by Rob Taviner

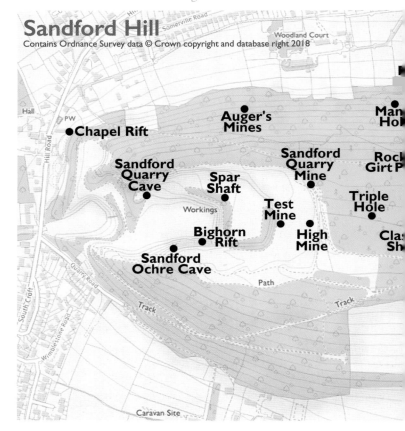

SAVILLE ROW SHAFT 1   Mine
ST 4303 5918   L 15m VR 15m Alt 113m

Another possible contender for Swarbrick-C-5, this 15m deep shaft, which is also known as MNRC-1 and SSSS-43-1, lies immediately adjacent to Whispering Gallery and is choked with woodland debris.

Ref: T. G. & J.T. Swarbrick, A Survey of the Lead, Ochre and Calamine Mines on Sandford Hill, p 61 (1948-1951), WNHAS & MNRC Report 46/47 pp 29-33 (1955/1956), SSSS, Sandford Survey Notes (1959-63) & E. Waters, Of Caves, Mines and Legendary Gulfs, SMCC Jnl 12 (1) pp 8-16 (2007)

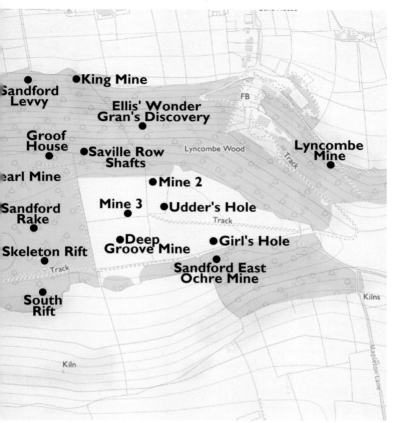

SHAFT 120   Mine
ST 4303 5918   L 95m VR 23m Alt 113m

Shaft 120 (also known as Swarbrick-C-4, Saville Row Shaft 4, MNRC-4 and SSSS-43-H), is a superb square-sided shaft, 23m deep, which intersects a mined gallery some 10m down and a second gallery situated just above the debris cone blocking the foot of the shaft. The Swarbrick report claims it to be 40m deep (probably incorrectly), and suggests there may be more galleries buried further down.

Ref: T. G. & J.T. Swarbrick, A Survey of the Lead, Ochre and Calamine Mines on Sandford Hill, pp 58-60 (1948-1951), WNHAS & MNRC Report 46/47 pp 29-33 (1955/1956), SSSS, Sandford Survey Notes (1959-63) & E. Waters, Of Caves, Mines and Legendary Gulfs, SMCC Jnl 12 (1) pp 8-16 (2007)

GROOF HOUSE SHAFTS Mine
ST 4298 5920 Alt 112m

This vast cone of spoil is peppered with choked shafts, indicating that mines of considerable depth once existed here. An attractive stone-lined pit, first recorded by the author in 2014 and located just below the north-west edge of the clearing, has been identified as a well-preserved Groof House - an 18th century miners' shelter. The MNRC report of 1955/56 records two open shafts in this location. The one furthest north, located in the clearing, was described as 7m deep to a small bell-shaped chamber, with a small passage leading towards the other open shaft located 10m further south in the wood. This second shaft was described as 20m deep and contained a rotten beam. However, these descriptions appear identical to Scum Hole and Rabbit Hole, which both the Swarbrick and SSSS surveys locate 100m further west. The most likely explanation is that there is a mistake in the MNRC report, but as all the shafts have now been filled in it is impossible to corroborate.

Ref: T. G. & J.T. Swarbrick, A Survey of the Lead, Ochre and Calamine Mines on Sandford Hill, pp 54-55. (1948-1951), P.A. E. Stewart, Report on investigations of mineshafts on Sandford Hill, WNHAS & MNRC Report 46/47 pp 29-33 (1955/1956) & SSSS, Sandford Survey Notes (1959-63)

SANDFORD GULF Of the many caves on Sandford Hill, none is as thought-provoking as the legendary lost 'Gulf', described in a letter penned by Rev. David Williams to John Rutter in 1829.

'I hope before you publish I shall be able to give you some account of an immense cave on Sandford Hill, which has never been explored, near which an elephant was found in 1770. The mouth of it is said by the miners to be 80 fathoms below the plane of the hill and they have let a man down upwards of 300 feet from its verge without coming to the floor, nor could he see any sides or termination to it - they call it the Gulph.'

A second letter, this time from Williams to Patterson, Rector of Shaftsbury, dated February 16th 1829, appears to correct the depth to 240 ft and add some details regarding the location of the elephant remains.

'The mouth of the largest which the miners call the Gulph lies they say 80 fathoms, or 480 feet below the plane of the hill. They also affirm that they have let down a man, with a line, 240 feet deep, but that he could see neither top, sides or bottom.' He went on to add ... *'There is another extensive cavern further to the westward in this hill, near which the skeleton of a full sized elephant was found in 1770'.*

The chasm was apparently discovered at some point between 1770 and 1829, when miners broke into a huge natural shaft into which a man was lowered until he became frightened out of his wits and called to be pulled back up. The entrance subsequently became blocked, possibly intentionally. Several contenders for the lost Gulf have been put forward over the years, including

Mangle Hole, Triple Hole and the Second Crossroads in Sandford Levvy, and there have even been suggestions that the lost cave may not exist at all, or that the term 'gulph', may merely have been a term used by miners for a large body of ore. However recent investigations by the MCRA have confirmed the Second Crossroads in Sandford Levvy to be by far the strongest contender. This theory was first formulated in Netherworld of Mendip (E. A. Baker & H. E. Balch 1907) who recorded that '*In driving an extensive level through a hill, 60 fathoms below the summit the miners came across a gigantic rift. A man was let down on a long rope but was unable to see walls or sides of this tremendous abyss.*'

Strangely, despite Sandford Levvy being the only extensive level on the hill, Baker and Balch appear never to have made the connection between the Levvy and Sandford Gulf, presumably because they never encountered any such feature. It was Willie Stanton (and others) who later realised the potential of the Great Rift at the Second Crossroads in Sandford Levvy, arguing that the rift originally continued down well below the current floor level and had subsequently been infilled with rubble extracted from the adit beyond. The diaries or mining journals of Captain William Webster would undoubtedly throw further light on its true whereabouts but sadly these have yet to be located.

Ref: The Sandford Gulf - D. Irwin, BEC Belfry Bull 426 p 3 (1984) & R. M. Taviner, The Sandford Gulf and other lost caves of Sandford Hill, (2016), available online via www.mcra.org.uk

FERN MINE   Mine
ST 4292 5919   L 6m VR 4m Alt 113m

Located in a grassy clearing and marked on the SMCC Survey as SMCC-6, this grilled 4m deep shaft yields a short horizontal passage ending in a collapse. It is also probably one of the sites mentioned in ACG Log Book No.2, p 42 (1968). It sits on a large area of spoil, and presumably once connected with the upper levels of the First Crossroads (east side) of Sandford Levvy.

Ref: C & C Dockrell, Sandford Hill Mines ACG N/L (Jan/Feb 1980) p2; E. Waters, Of Caves, Mines and Legendary Gulfs, SMCC Jnl 12 (1) pp 8-16 (2007)

SCUM HOLE   Mine
ST 4289 5916   L 20m VR 20m Alt 110m

Originally logged as Swarbrick-C-3, and later as MNRC-6 (but see Groof House Shafts), this 20m deep, cylindrical shaft was explored by SSSS in 1948-51, MNRC in 1956 & SSSS in 1959 but is now completely filled in. It is also marked on the Sidcot School Sandford Hill Speleological Survey of 1959 as SSSS-32-D. The shaft, which was 1-2m wide and cut through solid rock with no signs of walling, is probably one of the sites mentioned in ACG Log Book No.2, p 42 (1968). When first explored a piece of rotting timber was found wedged across the bottom of the shaft. A curious choked hole, located a few metres to the north, is probably the top of a ventilation shaft. This, and nearby Rabbit Hole, lie close to Hancock's Shaft, the uppermost point of the extensive First Crossroads (east side) workings in Sandford Levvy.

Ref: T. G. & J.T. Swarbrick, A Survey of the Lead, Ochre and Calamine Mines on Sandford Hill, pp 56-57 (1948-1951), P.A. E. Stewart, Report on investigations of mineshafts on Sandford Hill, WNHAS & MNRC Report 46/47 pp 29-33 (1955/1956) & SSSS, Sandford Survey Notes (1959-63)

RABBIT HOLE   Mine
ST 4288 5917   L 7m VR 7m Alt 109m

Also known as Egg Hole, this shaft has been logged as Swarbrick-C-2 and MNRC-5 (but see Groof House Shafts) and is marked on the Sidcot School 1959 survey as SSSS-32-C. Originally 7m deep to a small bell-shaped chamber, it is now completely filled.

Ref: T. G. & J.T. Swarbrick, A Survey of the Lead, Ochre and Calamine Mines on Sandford Hill, pp 54-55 (1948-1951), P.A. E. Stewart, Report on investigations of mineshafts on Sandford Hill, WNHAS & MNRC Report 46/47 pp 29-33 (1955/1956) & SSSS, Sandford Survey Notes (1959-63)

MCRA-SH-11   Mine
ST 4285 5916   Alt 109m

A choked mineshaft in a clearing bisected by the path. It lies 20m east of Pearl Mine.

PEARL MINE   Mine
ST 4285 5916   L 565m VR 30m Alt 109m

Also recorded as Sandford Pearl Mine, Sandford Hill Mine I, MNRC-7, SMCC-9, Swarbrick-C-1 and SSSS-32-A, this important mine was first explored by SSSS in 1949. The lower levels were blocked for several years until SSSS successfully reopened it in 1970 following work by ACG. It was blocked again in 1971 but was finally fully reopened by MCG and staff at Mendip Outdoor Pursuits following a dig commencing in 2016 when a significant number of new and important discoveries were made. The 20m deep rectangular entrance shaft intersects cross galleries which can be pursued at various levels to both east and west. The mine sports a host of interesting features including sharpening stones, stemple pockets, several animal skeletons, pick marks, smoked initials, drystone walls and stal-coated children's fingerprints. Several ascending shafts are encountered, some of which reach almost to choked pits on the surface, while a sound connection with passages above the 1st Crossroads in Sandford Levvy offers the potential for an entertaining future through trip. Formations can be found throughout, including the survivors of a famous nest of cave pearls, which Stanton described as being '... *the size of pigeon's eggs*'. It is now gated.

Ref: E. Waters, Pearl Mine Revisited, MCG News No.382 pp 8-17 (2018)

Pearl Mine. Photo by Andy Horeckyj

MCRA-SH-12
ST 4281 5916

Mine
Alt 109m

A choked mineshaft just to the north of the path. It lies 30m west of Pearl Mine and must lie close to Igloo Shaft in The Unexpected Level.

ROCK GIRT PIT
ST 4276 5916

Mine
L 3m VR 3m Alt 113m

Also logged as MCRA-SH-9, this deep, rocky pit, located next to the path, lies directly above (and very close to) the Girt Shafts near the western extremity of Pearl Mine.

COFFEE POT
ST 4276 5915

Mine
Alt 115m

A blocked mineshaft to the south of the track directly opposite Rock Girt Pit.

SSSS-22-B
ST 4274 5917

Mine
L 7m VR 7m Alt 113m

A 7m deep shaft marked on the Sidcot School Sandford Hill Speleological Survey of 1959. Located in a small grassy clearing immediately north of the path it has now been completely filled in. Along with SSSS-22-F, lies above the far western extremity of Pearl Mine.

Ref: SSSS, Sandford Survey Notes (1959-63)

SSSS-22-F
ST 4273 5916

Mine
L 7m VR 7m Alt 113m

A 7m deep shaft to the south of the track, marked on the Sidcot School Sandford Hill Speleological Survey of 1959. Situated in a line of deep depressions, it has now been completely filled in.

Ref: SSSS, Sandford Survey Notes (1959-63)

MCRA-SH-10
ST 4266 5920

Mines
Alt 105m

A small area of choked mineshafts.

SANDFORD QUARRY MINE
ST 4252 5915

Mine
L 17m VR 17m Alt 115m

Located by the track in the top rim of the quarry, this square vertical shaft was explored by ACG in 2004. The entrance is currently blocked with boulders (2014).

Ref: ACG Logbook No.9 p 30, 31, 33 & 37 (2004)

SUMMIT PATH

Follow the broad path which leads south up the hill from the ski centre car park. After 150m a junction is reached, where the entrances to Lyncombe Mine and Shaft can be found concealed in the undergrowth on the right. The remaining sites mostly lie on either side of the track which runs along the summit ridge.

LYNCOMBE MINE
ST 4355 5915

Mine
L 50m+ VR 10m Alt 77m

A small ochre working excavated on two levels. The 1m wide entrance slot yields a balcony where an awkward 4m pitch enters Blue Rabbit Chamber. A natural bridge leads to the upper and lower galleries. The mine was rediscovered by the BEC in 2003.

Ref: N. Harding, Lyncombe Mine, BEC Belfry Bull 53 (517) p 14 (2003)

LYNCOMBE SHAFT
ST 4352 5918

Mine
L 8m VR 8m Alt 85m

An 8m deep choked shaft on the surface above Lyncombe Mine.

Ref: M. Clarke, N. Gregory and A. Gray, Earth Colours – Mendip and Bristol Ochre Mining, MCRA, p 248 (2012)

SANDFORD EAST OCHRE MINE
ST 4331 5895

Mine
L 18m VR 18m Alt 117m

Also known as Star Ochre Mine, this WWII yellow ochre excavation was developed to help plug the gap caused by dwindling overseas supplies. Managed by the Golden Valley Ochre & Oxide Co., in collaboration with the Via Gellia Colour Co., it was a wide, waterworn pit, worked by three men from Derbyshire. The mine was serviced with a windlass and a rickety single-track narrow-gauge railway which carried the ochre across the south slope of the hill to a disused stone quarry adjoining Mapleton Lane. Here it was loaded into lorries for hauling to the Golden Valley works at Wick. The mine has now been completely filled in. The Swarbrick's noted several other small shafts on their survey in this area (Rift F), all of which appear to have been filled in.

Ref: T. G. & J. T. Swarbrick, A Survey of the Lead, Ochre and Calamine Mines on Sandford Hill, pp 81-90 (1948-1951) & M. Clarke, N. Gregory and A. Gray, Earth Colours – Mendip and Bristol Ochre Mining, MCRA, p 131 (2012)

GIRL'S HOLE
ST 4330 5900

Mine
L 10m VR 10m Alt 122m

Originally a large open pit with dangerously overhanging sides, this mine was named in honour of a girl from Sidcot who allegedly fell into it! The shaft yielded a small chamber from where a passage led east for 7m to terminate in a tiny chamber below a huge rock. When visited by SSSS c.1948, the mine was already in a parlous state and has now completely collapsed.

Ref: T. G. & J. T. Swarbrick, A Survey of the Lead, Ochre and Calamine Mines on Sandford Hill, pp 78-80 (1948-1951)

STANTON'S SHAFT
Mine
ST 4327 5899 L 10m VR 10m Alt 122m

A site explored by Stanton in 1950, who described '*a 10m deep mine shaft, 50m to the north-west of the big ochre pit.*' No sign of the mine now remains.

Ref: W. I. Stanton, Logbook 7 pp 21 (1950)

SSSS-52-D
Mine
ST 4321 5905 L 7m VR 7m Alt 123m

A 7m deep shaft marked on the Sidcot School Sandford Hill Speleological Survey of 1959. It has now been completely filled in.

Ref: SSSS, Sandford Survey Notes (1959-63)

UDDER'S HOLE
Mine
ST 4320 5907 L 40m VR 13m Alt 123m

Marked on the Sidcot School Sandford Hill Speleological Survey of 1959 as SSSS-42-J, this was described as a shaft into a reasonably-sized chamber. Daylight could be seen at the far end entering from a choked second entrance. Udder's Hole was filled in 1970-75 when the rake was levelled.

Ref: SSSS Logbook 5 p 26 (1949-1952) & SSSS, Sandford Survey Notes (1959-63)

SANDFORD HILL MINE 2
Mine
ST 4319 5911 L 40m VR 13m Alt 123m

Located at the eastern end of the northerly rake of Sandford Rifts East, a 13m deep shaft led to 25m of natural crystal-lined passage. It was filled with rubbish in 1970-75. This is the site listed as Swarbrick-D-6 by the brothers who regarded it as largely natural.

Ref: T. G. & J. T. Swarbrick, A Survey of the Lead, Ochre and Calamine Mines on Sandford Hill, pp 72-74 (1948-1951), W. I. Stanton, Logbook 6 pp 36-43 (1949-1950)

SANDFORD RIFTS EAST
Mine
ST 4317 5910 L 15m VR 6m Alt 123m

Two large, earth-choked natural rifts in a rake, partially excavated by SSSS in 1942. They were filled with rubbish in 1970-75 when the rake was levelled. The rifts were logged as Swarbrick-D-5 by the Swarbrick brothers who noted several similar but smaller rifts nearby and a small chamber at the western extremity.

Ref: SSSS Logbook 2 pp 23 & 29 (1941-1944), T. G. & J. T. Swarbrick, A Survey of the Lead, Ochre and Calamine Mines on Sandford Hill, pp 70-71 (1948-1951) & W. I. Stanton, Logbook 6 pp 36-43 (1949-1950)

SANDFORD HILL MINE 3
Mine
ST 4314 5906 L 20m VR 15m Alt 123m

Located in the southerly of two notable rakes, an 8m deep shaft led to a small collapsed chamber with shaft of 7m below. Deep rope-cut grooves in the entrance shaft indicated that it was once a far deeper working. It was filled with rubbish in 1970-75 when the rake was levelled.

Ref: W. I. Stanton, Logbook 6 pp 36-43 (1949-1950)

DEEP GROOVE MINE

ST 4312 5900

Mine
L 20m VR 15m Alt 123m

The mine was explored in the 1980s by Chris Richards (ACG) and Alan Dougherty (MCG) and subsequently obliterated when the field was levelled. A 13m deep mineshaft entered a roomy natural chamber, up to 7m high and 2m wide. A short extension to the west was choked with mud and rock and to the east a small passage choked after 7m. Both chokes probably connected with other shafts leading to the surface. The mine was formed along a fault displaying up to ten inches of fault breccia. As with nearby Sandford Hill Mine 3, deep rope-cut grooves in the entrance shaft indicated that it was once a far deeper working.

Ref: C. Richards, Personal communication

SOUTH RIFT

ST 4294 5889

Mine
L 15m VR 11m Alt 117m

A partially natural mine explored by ACG in 1972. A vertical pit leads to a steep rift, where a wriggle over a choke gains access to a short section of natural cave. The NGR quoted in the original ACG report is incorrect, which may have contributed to the same site later being 'rediscovered' and christened **Little Wood Mine**. The entrance is currently blocked with a pile of boulders (2014).

Ref: C. Richards, South Rift – a mined cave on the south side of Sandford Hill, Winscombe ACG N/L p 22 (July 1972) & Anon, Summit Plantation Mine, Little Wood Mine, ACG N/L pp 4-5 (July 1981)

SKELETON RIFT

ST 4293 5896

Mine
L 50m VR 35m Alt 125m

Also known as Skeleton Shaft or SSSS-32-P, the most reliable location for this mine is recorded in the Geoffrey and Swarbrick survey of (1948-51). When explored by SSSS in 1949 it consisted of a 35m deep, narrow, steeply-sloping shaft leading to a short section of natural passage which had been cleared of its earth infill. Later records showed it to be blocked at a depth of 3m and it now appears to have disappeared altogether. It was apparently named after an old gorse bush that was mistaken for a skeleton during the first descent! Unfortunately, the paths in this area of the hill have changed over the years and as all the shafts are now blocked it is impossible to identify the precise location of Skeleton Rift, Robert's Paradise or SSSS-32-N with absolute certainty.

Ref: T. G. & J.T. Swarbrick, A Survey of the Lead, Ochre and Calamine Mines on Sandford Hill, pp 76-77 (1948-1951), SSSS Logbook 5 pp 21-25 (1949-1952)

ROBERT'S PARADISE
Mine
ST 4292 5896 L 3m VR 3m Alt 124m
Marked on the Sidcot School Sandford Hill Speleological Survey of 1959 as SSSS-32-O, this shaft was once 3m deep, but has now been completely filled in.
Ref: SSSS, Sandford Survey Notes (1959-63)

SSSS-32-N
Mine
ST 4290 5896 L 7m VR 7m Alt 124m
A 7m deep shaft, marked on the Sidcot School Sandford Hill Speleological Survey of 1959. It has now been filled in.
Ref: SSSS, Sandford Survey Notes (1959-63)

SANDFORD RAKE
Rake
ST 4290 5904 Alt 125m
The sole remaining example of a series of large rakes excavated from west to east along the summit of the hill. It is aligned along a fault which may be connected with the Great Rift in Sandford Levvy, the probable location for the famous Gulf. There are a number of choked shafts that may repay detailed investigation.

CLASSIC SHAFT
Mine
ST 4281 5903 L 5m VR 5m Alt 125m
Logged as Swarbrick-D-4, this 1m diameter shaft is located in a grassy clearing 130m east of Triple Hole, and 60m to the north of the track which runs alongside the southern edge of the wood. It is covered with a grill and partially blocked with boulders.
Ref: C & C Dockrell, Sandford Hill Mines ACG N/L p 2 (Jan/Feb 1980)

SUMMIT PLANTATION MINE
Mine
ST 4277 5903 L 5m VR 5m Alt 125m
Located 100m east of Triple Hole and 60m to the north of the track which runs alongside the southern edge of the wood. A mine shaft, lined with dry stone walling reaches a small chamber. Below this, the shaft is blocked by a large boulder. It has now been completely filled in.
Ref: Anon, Summit Plantation Mine, Little Wood Mine, ACG N/L pp 4-5 (Jul 1981)

TRIPLE HOLE
Mine / Cave
ST 4268 5905 L 152m VR 37m Alt 125m
Also known as Sandford Rifts (West) and Triple Hole Cave, this is another site that has been put forward as a strong contender for both 'The Gulf' and, in particular, the lost Elephant Cave. The cave consists of three large rifts, interconnected and open to the surface. Although mostly natural they were undoubtedly exploited both for lead and for their ochre deposits, probably during the 18th and 19th centuries. The rifts unite above an

underground pitch below which several roomy chambers can be explored, littered with rubbish and (recent) bones. The lower cave was entered by CCG diggers in 1972 following excavation of the underground pitch. They found an 18th century clay pipe among the piles of deads, and a tally of ore scratched onto the wall of the main chamber. In 1995 a very large amount of animal carcasses and offal was dumped by persons unknown into the rifts and the National Rivers Authority poured over 300 litres of maggots onto the putrefying mass to hasten its removal. The Swarbrick brothers logged this site as Swarbrick-D-3 and called it 'The Grand Canyon' and recent investigations by the MCRA concluded that this was almost certainly the lost Elephant Cave (see below) and may also be the Sandford Bone Fissure excavated by Beard in 1833. Some of the Quaternary (Pleistocene) bones in the William Beard Collection in the Museum of Somerset at Taunton Castle, including wolf, lion, brown bear, woolly rhinoceros and spotted hyaena, may have originated from here.

Ref: Mendip Underground, MCRA pp 412-413 (2013)

Triple Hole. Photo by Rob Taviner

ELEPHANT CAVE This 'lost' bone cave, was described in a letter dated 30th January 1770 from a William Jefferies of Wrington to Dr Alexander Catcott, clergyman and geologist and author of A Treatise on the Deluge (1761 and 1768).

'According to your request I have procured all the Bones I mentioned which you will receive tomorrow; but while I was getting them up some particular parts were taken off although I was as careful as possible to preserve them; some of it I have recovered but one piece is still wanting which appears to be part of the Scapula but I think I shall be able to procure [word illegible here]. Where these Bones were found in almost the highest part of the Hill (Sandford hill, Somers.) on the north Side: they lay in an East and West Direction four fathom deep in a loose Strata composed mostly of small Fragments of limestone, Sand etc. and closely jammed in every way with Rocks; thence is a regular descent for about 100 yards 20 [word illegible] way round which induce one to think there is a Swallet near this Place, - I am still inform'd that there are more Bones left behind (but the workmen are so very imposing I shall not think of proceeding any further until I have had an opportunity of waiting on you to receive your Approbation (which shall be as soon as possible). I am now in haste to leave my Letter with the Carrier wch. I hope you'll admit us an Apology for all the Impropriation I have herein been guilty of and am. Your hbl. Servant. W. Jefferies Esq.

The discovery was summarised in Rutter's, The Delineations of North West Somerset (1829) in his description of Sandford Gulf.

'There is another extensive cave further to the westward, in this hill, near which, the skeleton of an elephant was found in 1770, four fathoms deep amongst loose rubble.'

H.E. Balch (Mendip – Its Swallet Caves and Rock Shelters p 100) tried to find this lost Elephant Cave, but failed, noting only that - *'At some distance [along the Levvy], a deep natural cavity which almost certainly reaches the surface, has been filled with compact cave earth. This might be worth digging, but I had no time to do so; it may contain important remains'* – an apparent reference to passages to the west of the First Crossroads in Sandford Levvy. Currant interpreted the location for this cave to be Triple Hole, and recent investigations by the MCRA arrived at much the same conclusion. The identification of the site is crucial, for it may act as a valuable pointer towards the location of 'Sandford Gulf (see separate entry)', historically described as lying close by, and to the east. The final resting place of the recovered elephant bones has never been discovered.

Ref: A. Currant, The Quaternary Mammal Collections at the Somerset County Museum, edited by C. J. Webster & R. M. Taviner, The Sandford Gulf and other lost caves of Sandford Hill, (2016), available online via www.mcra.org.uk

SANDFORD BONE FISSURE

Also known as Churchill Bone Cave and Sandford Hill Bone Fissure, this is the site opened by William Beard on 4th January 1838 in his search for bones. Named diggers included Robert Brown, William Cuff, Charles and John Venn while John Shepstone was paid a shilling for *'taking care of the cavern at Sandford Hill'*. Presumably the latter was employed in some capacity to guard the bone deposit. Entries for wages continued until the 29th May. Beard's collection, housed in the Museum of Somerset at Taunton Castle, contains one of the finest collections of cranial material in Britain, and at least some of the bones must have been derived from here. However Beard also suggests that he gave some of the bones to a Thomas Wright of Cheltenham. The site was apparently still open in 1863 when geologist James Parker applied to work at the site he described as *'an old mining shaft'*, but no records exist to show if he ever succeeded. It is not known if this was exactly the same site as Elephant Cave, although it seems very likely, especially as Beard regularly revisited the sites recorded by Catcott. Triple Hole is by far the strongest contender, but there are so many shafts, trenches and other mined features on Sandford Hill that it is impossible to say with any certainty.

Ref: The papers, notebooks and manuscripts of William Beard, held at the Somerset Record Office and James Parker, held at the Oxford University Geological Museum; The Lost Caves of Mendip - D. Irwin, BEC Belfry Bull 505 p 8 (1999) & R. M. Taviner, The Sandford Gulf and other lost caves of Sandford Hill, (2016), available online via www.mcra.org.uk

SSSS-22-O
Mine
ST 4265 5899 L 7m VR 7m Alt 127m

A 7m deep shaft, marked on the Sidcot School Sandford Hill Speleological Survey of 1959. It has now been completely filled in.

Ref: SSSS, Sandford Survey Notes (1959-63)

SSSS-22-N
Mine
ST 4262 5900 L 7m VR 7m Alt 127m

Another 7m deep shaft, marked on the Sidcot School Sandford Hill Speleological Survey of 1959. It has now been completely filled in.

Ref: SSSS, Sandford Survey Notes (1959-63)

SANDFORD HILL MINE 4
Mine
ST 4254 5903 L 36m VR 14m Alt 130m

A mine located on the very summit of the hill. Also known as SSSS-22-J, this site lay just east of High Mine, next to the summit trig point. One local legend suggested that the fabled lost Gulf lay somewhere nearby. A twisting 12m deep shaft led to a small chamber, where a further 2m drop reached a pool. A mined-out calcite vein could be followed east for 20m, while a blocked passage to the west probably once connected with High Mine. Sadly the mine has since been quarried away. This is probably the mine logged as Swarbrick-D-2, a site described by the brothers as containing ochre and having been *'left in an unfinished condition'*, possibly due to superstition surrounding a fatal accident.

Ref: T. G. & J.T. Swarbrick, A Survey of the Lead, Ochre and Calamine Mines on Sandford Hill, pp 63-64 (1948-1951), W. I. Stanton, Logbook 6 pp 36-43 (1949-1950)

HIGH MINE
Mine
ST 4253 5904
L 20m VR 15m Alt 130m

A mine located on the very summit of the hill, close to the trig point. Also known as SSSS-22-I, this is the site identified to both Stanton and Balch as being the traditional location for the legendary lost Gulf. A 14m deep shaft led only to a small collapsed chamber. Sadly the mine has since been quarried away, along with an associated large natural cavity filled with barren reddish clay that was exposed in the quarry face soon after its destruction.

Ref: W. I. Stanton, Logbook 1 p 13 (1943-1944) & Logbook 13 p 126 (1975) & R. M. Taviner, The Sandford Gulf and other lost caves of Sandford Hill, (2016), available online via www.mcra.org.uk

SSSS-22-K
Mine
ST 4252 5901
L 5m VR 5m Alt 130m

A 5m deep shaft, marked on the Sidcot School Sandford Hill Speleological Survey of 1959. Sadly the mine has since been quarried away.

Ref: SSSS, Sandford Survey Notes (1959-63)

TEST MINE
Mine
ST 4248 5905
L 9m VR 3m Alt 125m

This mine featured a large open shaft leading to a small mined chamber. The mine was quickly quarried away.

SPAR SHAFT
Mine
ST 4227 5897
L 17m VR 14m Alt 107m

This old ochre working was opened by quarrying in 1971. A 10m shaft led through calcite to a small chamber. The mine was quickly quarried away, although the calcite vein can still be seen in the north wall of the quarry.

Ref: C. Richards, Spar Shaft – a new mine on Sandford Hill, ACG N/L pp 51-52 (Oct 1971)

SANDFORD QUARRY CAVE
Cave
ST 422 591
L 21m VR 8m Alt 76m

Also known as Sandford Hill Quarry Cave it was opened by quarrying in 1942. The cave comprised a large chamber with several passages leading off. It was filled with rubble within a week. This was only one of several cavities discovered and subsequently destroyed during active quarrying operations, and some large ochre-filled cavities still remain visible.

Ref: SSSS Logbook 2 pp 27-29 (1941-1944)

SANDFORD QUARRY BADGER CAVES Cave
ST 4213 5914 Alt 75m

Two small badger-infested holes first recorded by ACG in 1991. They are in the same vicinity as Sandford Quarry Cave and there is a nice crystal pocket and another high level badger-sized hole in the cliff further west.

SANDFORD QUARRY CAVE 2 Cave
ST 4229 5902 L 23m VR 7m Alt 75m

Best reached by a 5m climb from below, a roomy descent of 5m yields a small, steeply-ascending, tight and loose bedding cave. It was exposed in the southern wall of the upper quarry in 1968.

Ref: J. Smart, Sandford Quarry Cave No.2, CCG N/L 4 p 18 (Oct 1968 – Mar 1969)

BIGHORN RIFT Cave
ST 4230 5900 L 30m VR 10m Alt 100m

Located close to the top of the south face this is a high scalloped rift with a phreatic tube in the roof which terminates in a collapse. It was known to climbers but was first officially recorded by MCG in 2017.

Ref: M. Moxon, Bighorn Rift, MCG News No.383 p 23 Sep (2018)

SANDFORD OCHRE CAVE Cave
ST 4227 5899 L 22m VR 6m Alt 73m

Marked on the Sidcot School Sandford Hill Speleological Survey of 1959 as SSSS-II-A and also known as Sandford Ochre Mine and Three Entrance Mine. A cave with three entrances, including one in the roof. Other than Sandford East Ochre Mine, it was the last active pit on Sandford Hill and was excavated by ochre miners in 1930-40. Notable for being Willie Stanton's 'first cave', it was buried under tipped overburden in 1974.

Ref: W. I. Stanton, Logbook 1 pp 1-2 (1941)

SSSS-02-A Mine
ST 4211 5919 L 2m VR 2m Alt 90m

A 2m deep shaft near the edge of the old quarry. It can no longer be identified.

Ref: SSSS, Sandford Survey Notes (1959-63)

CHURCHILL

The village of Churchill lies at the intersection of two major turnpikes (the modern day A38, also known as the New Bristol Road, and the A368), and was once guarded by the Churchill Gate, a notable two-storey hexagonal toll house complete with Regency veranda. Sadly, this was pulled down in 1961, when the junction was upgraded. The New Bristol Road was constructed by out-of-work miners and paid for by monies granted in an 1825 Act of Parliament. It bypassed the old turnpike road which ran along the top of Dinghurst Camp to the west, although the old road still exists as a track which runs in front of the Crown Inn. The famous Churchill family, whose lineage includes such notable luminaries as the Duke of Marlborough and Sir Winston Churchill, actually derives its name from this parish, which was home to an ancestor who settled here after the Norman Conquest. On 18th November 1940, American magazine *Life*, published a major wartime feature on the village and its residents. Called '*Churchill, England, What We Fight For*', it included many photographs taken by Cecil Beaton. Churchill himself even penned a commendation, intimating that the images gave American readers a much better idea of the way of life Britain was fighting to defend.

CROWN INN SWALLET
ST 4455 5952
Cave / Sink
L 15m VR 3m Alt 75m

Also known as Crown Inn Cave and Churchill Swallet, this interesting fenced-off 'sink' is located by the side of track 100m uphill of the Crown Inn. Now almost full to the brim with debris, it swallows the roadside drainage which has been postulated may feed Banwell Spring. In 1995, storm water could be heard running away freely along a descending tunnel. There was also a natural chamber nearby with small offshoots, which had been worked for ochre.

Ref: C & C Dockrell, Additions to the Complete Caves of Mendip, ACG N/L pp 1-2 (Jun 1992) & N. Richards, Personal communication

HEAL'S OCHRE MINES
ST 4454 5947
Mine
L 12m VR 4m Alt 81m

Between 1906 and 1920, Richard and Owen Heal worked ochre in this vicinity. Rowberrow residents, their best documented working lay alongside the track, near the spot marked on old maps as Victoria Cottages (now demolished) and was entered by two shafts, one situated in the adjacent field. A fissure here apparently ran across the track from east to west, and the eastern part is said to have intersected some old workings alongside the A38. The Heal's worked the western end and a windlass remained in situ until the 1920s. Part of this mine was rediscovered by '*The Two Nick's*' (Harding & Richards) in 2009, following a road collapse. A 4m deep shaft yielded a steeply-sloping passage to a miners' wall and blocked second shaft. The collapse was swiftly filled in. Given the dates and location, Heal's workings must be a strong contender for the Lost Cave of Churchill (see below). The Heal's also worked the northern slopes of Dolebury Warren and a cluster of pits at ST 4988 5913 are presumably remnants of their handiwork.

Ref: West Mendip Update, Descent 212 p 10 (Feb/Mar 2010), M. Clarke, N. Gregory and A. Gray, Earth Colours – Mendip and Bristol Ochre Mining, MCRA, p 130 (2012) & P. Burr, Mines and Minerals of the Mendip Hills, Volume II, MCRA p 565 (2015)

LOST CAVE OF CHURCHILL   Lost Cave

This lost cave was said to have been explored 'about 1900' and was described in the Women's Institutes 'The Story of Churchill with Langford'.

'I have heard my father tell of the existence of an underground cave and channel with an entrance (since lost) somewhere in a field on top of Churchill Batch. His father with a Mr Baker, of the old blacksmith family, and others explored this cave with candles. They traversed a considerable distance underground until they were stopped by an underground sheet of water. They calculated they were the under Windmill Hill, close to the church and formed the conviction that Bishop's Well, opposite the Church, which breaks after a prolonged wet spell, is an overflow of this underground lake.'

Numerous searches for the lost cave have been conducted, included one by the ACG in the 1980s which lasted for several weeks. Several 'Batches' are recorded on maps of the area. Modern maps locate 'Churchill Batch' above Churchill Quarry to the east of Dinghurst Camp (even east of the A38), and there is another 'Old Batch' to the south of Knowle Wood where some mining certainly took place (see Knowle Wood Mines below). However, the one referred to is generally accepted to be the hillside above the Crown Inn, through which the Old Coach Road runs and where Crown Inn Swallet and Heal's Ochre Mines are located. Before the present route of the A38 was cut through the hill at Churchill Rocks in 1819, this forgotten little lane was actually the main turnpike between Bristol and Bridgwater. There is no mention in the account of how the cave was first discovered. Quarrying in the vicinity was in operation well before 1900 (**both Heal's are listed as quarrymen on census returns of the period**), and caused considerable damage to the Iron Age hillfort of Dinghurst Camp, whose woody hollows are still clearly visible to the east of the track. Mining too probably occurred well before the Heal's sinking shafts c.1906. However, given the proximity of 'about 1900' to the date the Heal's began operations, the most likely site for the lost cave is in one of their workings – possibly even the one examined by Harding & Richards in 2009, while the '*underground sheet of water*', may relate to the water which now sinks at Crown Inn Swallet. Richards also believes that the cave lies somewhere in this vicinity and was even shown an old dig from 1947 which was sunk during an attempt to locate it. Two ponds were revealed! The association with Windmill Hill and the

Bishop's Well is clearly a gross exaggeration. Both lie roughly 1km to the north of Churchill Batch and are separated from the main mass of the hill by low-lying ground. There is indeed a spring by the Parish Church called **Bishop's Well** (ST 4378 6026), and another close by called **The Bridewell** (ST 4409 6047), where a newlywed bride is said to have drowned herself on her wedding day. Notwithstanding that a far more likely explanation is that the name is simply a corruption of the Old English for Bird Well, the bride's ghost seems determined to have her day and is said to appear still dressed in her 'shimmering' bridal gown. This shimmering effect seems to be a common occurrence near water and it is interesting to speculate that it may be a form of mirage, the result of light refraction caused by differences in temperature between the air immediately surrounding the water and the layers above it.

Ref: The Story of Churchill with Langford, The Churchill's Women's Institute & N. Richards, Personal communication

KNOWLE WOOD MINE
Mine
ST 4387 5926
L 21m VR 11m Alt 70m

Also known as Knowle Mine and Town Pit, this is the largest surviving working in Knowle Wood. A fenced, 6m deep shaft, yields a roomy eastern gallery which extends for 7m to a dead end where pick marks are much in evidence. To the west, a much smaller gallery contains stacked deads and numerous phreatic solution hollows. A prominent calcite vein can be traced throughout the length of the working and the miners probably followed this in search of lead or ochre.

Ref: Mine Sites on Churchill Knowle, N. Richards & N. Harding, BEC Belfry Bull 520 pp 11-12 (2004)

CALCITE SHAFT
Mine
ST 4390 5922
L 6m VR 6m Alt 90m

This fenced mineshaft lies at the intersection of two massive calcite veins and was probably dug in search of lead ore.

Ref: Mine Sites on Churchill Knowle, N. Richards & N. Harding, BEC Belfry Bull 520 pp 11-12 (2004)

KNOWLE CAVE
Mine
ST 4392 5921
L 7m VR 2m Alt 100m

Located immediately above Calcite Shaft and close to a mysterious fenced tree, this is a large open pit with a roomy natural rift beneath the north wall, formed along a massive calcite vein. The rift, which was re-entered by the BEC in 2002, is separated from the main pit by a thin skin of bedrock which had been breached by ochre miners. The pit is now largely filled with leaf mould and the rift is no longer accessible.

Ref: Mine Sites on Churchill Knowle, N. Richards & N. Harding, BEC Belfry Bull 520 pp 11-12 (2004)

WINSCOMBE

The low-lying village of Winscombe, lies in-between the two arms of the west Mendip range, in the Lox Yeo valley. Although there are no caves of any note, lead and calamine were both once mined here, and several strong springs emerge from the underlying Dolomitic Conglomerate. In 2009, the village became the subject of 'The Winscombe Project', an important historical and archaeological study led by Professor Mick Aston of Time Team fame, who published the results in a series of articles in the *Proceedings of the Somerset Archaeological and Natural History Society*.

MAX MILL
Mill Pond
ST 4029 5758
Alt 15m

A mill has stood on this site for centuries. What is less certain is whether or not it was ever spring fed. There have been some suggestions that originally it was at least partially sourced by upwellings in the bed of the mill pond, which dried up when pumping started at Cox's Well. However there are no historic records to support this, and this suggestion appears to have become confused with nearby Five Springs, which rise in the bed of Winscombe's other mediaeval mill pond, and where the exact same thing is said to have happened. Instead the mill appears to have always been sourced by a 500m long, man-made leat, which gathers its waters from many of the district's other healthy springs, including Cox's Well, Five Springs, East Well, Hale Well and **Fidling's Well / Fidling Wells** (ST 4116 5720).

Ref: M. Aston et al, Medieval Settlements in Winscombe Parish in North Somerset, SANHS Proceedings 153 pp 55-90 (2010)

COX'S WELL
Spring
ST 4083 5772
L 34m VR 34m Alt 19m

The site of a strong spring rising from Red Marl above Dolomitic Conglomerate, the source was capped in 1887, to help provide a reliable water supply to the town of Burnham-on-Sea. To augment this supply, four boreholes were later sunk around the utilitarian brick-built structure, and the increased pumping is colloquially claimed to have reduced the flow from nearby Five Springs. Prior to being capped, the spring was said to be the lure of an otherworldly ghoul.

Ref: L. Richardson, Wells and Springs of Somerset, HMSO pp 163-164 (1928) & N. Barrington, W. Stanton, The Complete Caves and a view of the hills, p 58 (1977)

FIVE SPRINGS
Spring
ST 4129 5751
Alt 20m

Knight, in his Heart of Mendip, recorded '*a cluster of powerful springs*', rising through Red Marl and Dolomitic Conglomerate, in the bed of a mediaeval millpond. The ruins of Woodborough Mill are still visible a short distance downstream. The flow is claimed to have reduced when pumping began at Cox's Well.

Ref: N. Barrington, W. Stanton, The Complete Caves and a view of the hills, p 78 (1977)

EAST WELL
ST 4148 5667

Spring
Alt 40m

Arguably the most interesting of Winscombe's many springs, East Well is the perennial source of Winscombe Brook. Surrounded by the remains of old masonry, it rises in Red Marl overlying Old Red Sandstone, and bears all the hallmarks of a holy well, while its relative position, as the nearest natural water source to the mediaeval church of St James, strongly suggests that its waters may once have been used for baptisms. It is also known locally as the Ice Well, on account of it never freezing over! There are other smaller springs in the vicinity and the remains of an ancient stone footbridge can be seen nearby.

Ref: P. Quinn, The Holy Wells of Bath and Bristol Region, Logaston: Almeley, p 200 (1999)

WOODBOROUGH GREEN
ST 4200 5750

Mines
Alt 30m

In 1650, no less than 100 tons of lead ore were extracted from Woodborough Green during a three month period, and 17th century maps show a line of calamine pits in the vicinity of the quoted NGR. The mines were said to be very rich, and John Bettington, a Bristol lead merchant, tried to exploit this advantage, continuing operations here in the late 18th and early 19th centuries. All signs have now been lost beneath housing.

Ref: M. Clarke, N. Gregory and A. Gray, Earth Colours – Mendip and Bristol Ochre Mining, MCRA, p 156 (2012)

SIDCOT SCHOOL SHAFT
ST 4289 5744

Mine
Alt 80m

In 1863, a violent storm uprooted a tall elm tree growing at the eastern end of the school terrace. Investigations found that it had been growing inside the mouth of an old mine shaft, which presumably was swiftly filled in.

Ref: F.A. Knight, A History of Sidcot School – A Hundred Years of West Country Quaker Education 1808-1908, J. M. Dent & Company, p 200 (1908)

SIDCOT SCHOOL WELL
ST 4287 5750

Well
Alt 83m

In 1809, a well sunk in *'the south-west corner of the boy's playground'*, intercepted a strong spring at a depth of *'about 400 feet'*, which provided the school with an independent supply of reliable clean water. This served the school well until 1820, when a son of Charles Strode, an employee of the school, was accidentally killed while operating the machinery.

Ref: F.A. Knight, A History of Sidcot School – A Hundred Years of West Country Quaker Education 1808-1908, J. M. Dent & Company, pp 66-67 (1908)

FULLER'S POND
ST 4246 5693

Spring
Alt 57m

This small, spring fed pond, was apparently once much larger, and was a popular local skating rink for Sidcot schoolboys. Following the unfortunate accident, which closed the school well in 1820, the school took water from this pond to use for washing.

Ref: F.A. Knight, A History of Sidcot School – A Hundred Years of West Country Quaker Education 1808-1908, J. M. Dent & Company. p 68 (1908) & L. Richardson, Wells and Springs of Somerset, HMSO p 47 (1928)

HALE WELL

Spring

ST 4276 5665

Alt 90m

Also recorded as Hales Well, this timid spring, which rises in a small stone trackside tank, was Sidcot School's original clean water supply and was a favourite haunt of their scholars, including F.A. Knight, the author of Rambles of a Dominie, who immortalized it in his account of 'Sleepy Hollow.' *It is an ancient spring. The hands that fitted these broad flagstones round its brink were folded for their last sleep long years ago. For centuries the sons of toil have cooled their sunburned faces in a well that never in the memory of man has failed or faltered in its flow. The hottest summer never checked its bounteous stream; the keenest winter never laid a curb upon its freedom.* Small amounts of both iron and yellow ochre are said to have been raised from workings near Oakridge Farm on Sidcot Hill, and around Hale Combe.

Ref: F.A. Knight, The Rambles of a Dominie, Wills Gardner, 1891 & A History of Sidcot School – A Hundred Years of West Country Quaker Education, 1808-1908

WINSCOMBE RAILWAY CUTTING CAVES

Caves

ST 4218 5597

L 2m VR 1m Alt 40m

There are a handful of tiny holes visible in the deep cutting approaching the southern end of the 165m long Shute Shelve tunnel. The most obvious, located at track bed level, 55m south of the tunnel entrance, is a narrow stal-encrusted rift which was dug briefly by ACG in 1970. In warm weather, a cool draught is noticeable, probably connected with a small square-shaped hole located a few metres above. There are other small holes in this vicinity, while on the opposite bank at least two impenetrable high level rifts are visible, one situated directly opposite the ACG dig, and another much closer to the tunnel mouth. The Cheddar Valley Railway (better known as the Strawberry Line), opened in 1869, and operated until 1963. Part of it has been transformed into a 10-mile long, popular cycle path. Before it was much diminished by road widening, Shute Shelf was once quite a dramatic local landmark, and old paintings and photographs clearly show a distinctive cliff, topped with brooding, malevolent trees, which the infamous Judge Jeffreys used as gibbets during his Bloody Assizes.

Ref: N. Barrington, W. Stanton, The Complete Caves and a view of the hills, p 176 (1977)

STAR & SHIPHAM

'But the loveliness of the weather did not fail, and the whole day was set in Severn landscapes. They first saw the great river like a sea with the Welsh mountains hanging in the sky behind as they came over the Mendip crest above Shipham'.

Notwithstanding the flowery prose of H.G. Wells, Shipham has always been synonymous with mining. Rich deposits of zinc and lead are found hereabouts and in the 18th and 19th centuries, calamine (zinc carbonate), blende (zinc sulphide) and galena (lead sulphide), were all extracted in huge quantities. Indeed, calamine extraction began here as early as the 16th century, and was worked more or less continuously for more than 250 years. The shafts were often small-scale ventures, operated by family groups, and Collinson, writing in 1791, noted more than 100 shafts in operation, some up to 22m deep. The men and the older boys worked below, while the women of the family, and the smaller children, controlled the surface, hauling and washing the ore. Billingsley estimated that up to 500 shafts had been sunk in the Shipham and Rowberrow region alone, and even when deposits were later found elsewhere in the United Kingdom, the ore extracted from Shipham was still regarded as the best. The largest of the mines (Singing River Mine), contains a stream and was used as an underground reservoir in the 1920s. One 19th century calamine processor still exists, and is a Grade-II listed building and this rich mining history is reflected in some of the street names. The village was once a notorious no-go area for civilised folk and it is said that no self-respecting policeman would dare try and arrest any of the local men, lest they end up at the bottom of one of the localities numerous pits. The miners' were also quick to war, and Shipham men played prominent roles both in defeating the Spanish Armada, and in the Monmouth Rebellion. The village was eventually 'tamed' by Hannah More (1745-1833), a famous playwright and philanthropist and acquaintance of such luminaries as David Garrick, Dr Johnson and Sir Joshua Reynolds. A passionate anti-slavery campaigner, this original 'Bluestocking' was encouraged by William Wilberforce to help civilise villagers brought low by a collapse in the price of calamine, who he described as *'savage and depraved almost beyond Cheddar'*. Despite considerable initial local opposition, Hannah, together with her sister, Martha, successfully established a special school, one of a dozen purchased in the area to help deliver a basic education to the masses. By cornering the market in local calamine production, Hannah even managed to win over the Shipham miners, using her influence and connections to negotiate a much fairer price on their behalf with the all-powerful brass merchants in Bristol. Despite the importance of mining to the village, Shipham's most famous resident was probably Smithson Tennant (1761-1815), who built Longbottom Farm, and was the first to identify the elements osmium and iridium, both by-products of platinum production. His business partner later identified rhodium and palladium and as a result of their success Tennant went on to become a Fellow of the Royal Society and Professor of Chemistry at Cambridge. Sadly he was killed in a freak accident, when his horse fell into a trench in France. The mineral tennantite is named after him. During WWII, the village played host to a searchlight battery and a listening post, whose main job was to try and bend the radio beams guiding the German bombers so that they dropped their payloads onto decoy 'Starfish' sites, rather than their intended targets. There were several Starfish sites dotted around Mendip, including the original on the top of Blackdown, where an entire replica of Bristol, complete with streets and railway lines, was reconstructed using the help of Shepperton Film Studios. An elaborate and largely successful decoy, it consisted of dimmable light bulbs mounted on piles of stones, controlled from blast-proof bunkers which remain visible today.

STAR

PYLE WELL
ST 4348 5859

Spring
Alt 63m

Also known as Pylle Well, this covered spring rises in swampy ground overlying Dolomitic Conglomerate. The summer source of Towerhead Brook, it lies just outside the enclosure of a Roman villa, and close to the Wimblestone, an impressive standing stone measuring almost 2m high and over 1.5 m wide. There may once have been other stones nearby. Local legend tells that that the stone can go 'walkabout' during the dead of night, revealing 'a heap of shining gold' secreted underneath. Apparently one local farmer even tried moving the giant stone to retrieve it, but not surprisingly, was defeated.

Ref: N. Barrington, W. Stanton, The Complete Caves and a view of the hills, p 125 (1977) & L. Richardson, Wells and Springs of Somerset, HMSO pp 46-47 (1928)

HAM MINE 1
ST 4399 5852

Mine
L 27m VR 12m Alt 80m

This interesting small mine was one of three uncovered c.1987 by workmen laying a new water pipe along The Batch. Located in the front garden of Laneside, a 9m deep, steeply-sloping shaft yielded a small upper gallery and a further 6m drop into a lower gallery. It was named after John Ham, the then owner of the cottage and stalwart of both NHASA and WCC.

Ref: P. G. Hendy, Star Mine and Related Sites Part 2, WCC Jnl 27 (290) pp 163-165 (2004)

HAM MINE 2
ST 4398 5850

Mine
L 10m VR 3m Alt 80m

Located adjacent to the telegraph pole, a short shaft yielded a narrow passage running parallel with, and below, the lane. The workmen intercepted a third mine (**Ham Mine 3**, ST 4399 5851) which was capped before it could be explored.

Ref: P. G. Hendy, Star Mine and Related Sites Part 2, WCC Jnl 27 (290) pp 163-165 (2004)

PARADISE VALLEY

Paradise Valley is the rather grandiose name given to the narrow defile which connects the villages of Star and Shipham. The northern flank in particular has been extensively mined, and several small localised workings remain penetrable. Immediately above the valley, mining took place on a much grander scale, especially at Star Shaft, an impressive Cornish shaft sunk in the mid-19th century. During the 1940s, Paradise Valley became the private playground of the Sidcot schoolboys, who explored all the mines open at the time and gave them names which are no longer in use. However, where known they have been supplied as alternatives. The sites are described in a downhill direction, starting from the new village hall and returning via Star Common. Several are inhabited by bats and thus should be avoided during the hibernation season.

NEW HALL MINE
Mine
ST 4443 5787
L 12m VR 8m Alt 133m

Discovered during construction of the new village hall in 2005, this locked shaft lies alongside the path which runs behind the building close to a lamp post. Mined for both lead and calamine, a lack of shot holes suggests that it may be one of the oldest mines explored in Shipham to date. The shaft yields short workings aligned along an east-west axis.

Ref: R. G. Witcombe, Who Was Aveline Anyway? – Mendip's Cave Names Explained. WCC Occasional Publication (2008)

DGW MINE
Mine
ST 4438 5790
L 11m VR 9m Alt 130m

Named after David George Wilson Hain, a relative of Mark Ireland who explored it in 2005. A natural rift in a gruff apparently mined for calamine, it has since largely been filled in. In 2016 only the top couple of metres remained visible beneath a pallet and some corrugated iron.

Ref: R. G. Witcombe, Who Was Aveline Anyway? – Mendip's Cave Names Explained. WCC Occasional Publication (2008)

RUBBISHY RIFT
Mine
ST 4430 5795
L 3m VR 2m Alt 124m

This site lies just south of the path that runs alongside the football pitch. Now largely filled in, it was dug briefly by the ACG in 1971.

Ref: C. Richards, Descriptive list of the open mines of Shipham, ACG N/L pp 3-6 (Jan 1972)

TIN CAN ALLEY
Mine
ST 4428 5796
L 60m VR 20m Alt 120m

Referred to earlier as Daffodil Valley Cave, this mine is located on the northern slope of the valley immediately above Daffodil Mine and is easily the most visually impressive of the sites still penetrable in Paradise Valley. A steep muddy

New Hall Mine. Photo by Alison Moody

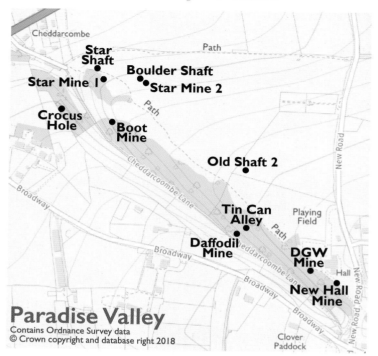

slope yields a small chamber heavily fouled with rubbish, with a continuation which doubles back beneath the entrance to reach a 6m deep pitch. This yields a series of galleries approaching 40m in length, probably heading towards several blocked shafts which can be seen further down-valley at the same level.

Ref: P. North, Speleological Diary pp 101-103 (1929-30)

DAFFODIL MINE

Mine

ST 4427 5794
L 77m VR 20m Alt 115m

A good surviving example of a small-scale calamine and lead working, this shaft, located immediately above the path that runs along the valley floor, is covered with rusty metal and surrounded by bright orange tape. A 5m deep shaft leads to a further 8m descent down a wall of deads. This gains additional workings at both higher and lower levels with some evidence of natural cave passage. An alternate and deeper entrance shaft once existed just to the east, but this is now blocked with rubbish. Although it was probably known to the Sidcot schoolboys, the first recorded exploration was carried out by ACG in 1971.

Ref: C. Richards, Daffodil Mine and Crocus Hole, Shipham, ACG N/L pp 93-97 (Dec 1971) & Mendip Underground, MCRA p 107 (2013)

BOOT MINE
Mine
ST 4403 5818 L 6m VR 4m Alt 102m

Also known as New Shaft 2, a 2m drop leads via a steep descent to a small mined rift with a choked shaft in the floor This is probably the arrestingly named 'Sambo' (Willie Stanton's schoolboy nickname), first explored by SSSS in 1946. There is a small open gruff nearby which has been christened the **T-Shaped Gruff** for obvious reasons.

Ref: P. G. Hendy, Star Mine and Related Sites Part I, WCC Jnl 27 (289) pp 143-145 (2004) & SSSS Logbook 3 p 65 (1946)

HAZEL MINE
Mine
ST 4402 5819 L 5m VR 5m Alt 98m

Also known as New Shaft 1. A 3m drop leads into a walled rift that is partly capped. A steep descent over leaf mould leads via a squeeze to a mud floor.

Ref: P. G. Hendy, Star Mine and Related Sites Part I, WCC Jnl 27 (289) pp 143-145 (2004)

NEW SHAFT
Mine
ST 4402 5819 L 3m VR 3m Alt 97m

The shaft lies at the edge of the wood, immediately above the opencast ground containing Holly Fissure. A narrow slot yields a choked rift.

HOLLY FISSURE
Mine
ST 4401 5819 L 6m VR 5m Alt 95m

This small mined vertical rift is located in a gruff at the south-east limit of an area of opencast workings in Paradise Valley, just below New Shaft. The entrance is covered with corrugated iron.

Ref: C. Richards, Descriptive list of the open mines of Shipham, ACG N/L pp 3-6 (Jan 1972)

IVY FISSURE
Mine
ST 4401 5820 L 3m VR 2m Alt 97m

This small mined rift at the bottom end of Paradise Valley lies in a bramble and rubbish choked pit below an old pulley (still present 2016). A small drop yields a short crawl.

Ref: C. Richards, Descriptive list of the open mines of Shipham, Shipham, ACG N/L pp 3-6 (Jan 1972)

CROCUS HOLE
Mine
ST 4395 5819 L 27m VR 10m Alt 101m

This attractive, rectangular, 9m deep shaft is the only one located on the southern flank of the valley. It lies next to the wall at the top of the wood and is covered with a metal grill. The shaft yields workings trending in a westward direction. The main rift leads to a boulder choke, while a second passage leads south through a crawl to a choked chamber.

Ref: C. Richards, Daffodil Mine and Crocus Hole, Shipham, ACG N/L pp 93-97 (Dec 1971)

STAR COMMON

STAR SHAFT
Mine
ST 4401 5827 L 163m VR 46m Alt 97m

Also known as the Rowberrow Shaft, this very impressive Cornish shaft was sunk in the mid-19th century (1865-1870). The gate yields a 28m deep rectangular shaft (originally 73m deep), which passes through some (unexplored) 'old men's workings', which clearly pre-date the Cornish work. A short passage then reaches a further shaft to a lower level, which contains some important tools and artefacts that should be left in situ. This level is often flooded, and on occasions water even backs up to the bottom of the entrance shaft. The shaft was uncovered in 2001 when a tractor driver nearly lost his vehicle down it. Subsequently entered by WCC, the flooded lower workings were explored by CDG. The remains of cataract and hot pits associated with a Cornish beam engine have been excavated in the woods near the entrance, along with the foundations of a boiler house.

Ref: Mendip Underground, MCRA pp 330-333 (2013) & P. Burr, Mines and Minerals of the Mendip Hills, Volume I, MCRA pp 322-326 (2015)

STAR MINE 1
Mine
ST 4402 5824 L 24m VR 9m Alt 102m

Also known as Harrow Pit, Mine One and Shaft One, this grilled mineshaft lies in open gruffy ground about 30m south-east of Star Shaft. A 9m deep shaft leads to three radiating passages.

Ref: C. Richards, Two mines near Star, Shipham, ACG N/L pp 63-65 (Aug/Sep 1971), N. Watson, Star Mines, Shipham, BEC Belfry Bull 33 (376/7) p 18 (1979) & P. G. Hendy, Star Mine and Related Sites Part 2, WCC Jnl 27 (290) pp 163-165 (2004)

BLIND PIT
Mine
ST 4403 5822 L 1m VR 1m Alt 103m

This choked shaft, situated on the edge of the common, is covered in corrugated iron. It may be the mine christened Arsey Pit 1 by Sidcot Schoolboys in 1946 and described as 13m deep with a short side passage. There are many other deep pits nearby, mostly overgrown with brambles, where another loose and dangerous 13m deep shaft with a 7m side extension known as **Wallaby Hole** was also explored by SSSS in 1946.

Ref: P. G. Hendy, Star Mine and Related Sites Part 1, WCC Jnl 27 (289) pp 143-145 (2004) & SSSS Logbook 3 p 65 (1946)

STANTON SHAFT
Mine
ST 4407 5826 L 50m VR 13m Alt 99m

In 1946, SSSS descended a 17m deep shaft in this vicinity after belaying a ladder to a large anthill! The shaft yielded a mined-out calcite vein which extended for more than 25m in one direction and 7m in the other. This does not appear to be Boulder Shaft, although the given location is very similar. The quoted NGR marks the most likely site - a blocked shaft filled with metal and decomposing tyres.

Ref: SSSS Logbook 3 pp 65-66 (1946)

BOULDER SHAFT
Mine
ST 4408 5826
L 40m VR 10m Alt 99m

Also known as Shaft Two, this mine has been capped with large rocks above corrugated iron. A 10m descent yields a rising level which terminates in a choke. There is another passage which rejoins the shaft 3m above the floor behind a wall of deads, opposite another level which terminates in a small choked aven.

Ref: N. Watson, Star Mines, Shipham, BEC Belfry Bull 33 (376/7) p 18 (1979) & A. Gray Logbook pp 5-7 (2006)

STAR MINE 2
Mine
ST 4409 5826
L 9m VR 3m Alt 101m

A short drop leads to a short muddy crawl in a mined rift which is now choked but once led via a squeeze into an ascending rift 6m long. Partially covered with a small iron frame, this is almost certainly the mine originally christened Arsey Pit II by SSSS in 1946. It was filled in but re-discovered by ACG in 2006, who called it Bedstead Rift.

Ref: C. Richards, Two mines near Star, Shipham, ACG N/L pp 63-65 (Aug/Sep 1971), SSSS Logbook 3 p 65 (1946) & New Finds / Re-discoveries, A. Gray, ACG Jnl pp 27-33 (2006)

COLLARED PIT
Mine
ST 4407 5823
Alt 105m

This blocked pit on the common is surrounded by a ring of spoil. There are many other deep pits nearby, mostly overgrown with brambles.

Ref: P. G. Hendy, Star Mine and Related Sites Part 2, WCC Jnl 27 (290) pp 163-165 (2004) & SSSS Logbook 3 p 65 (1946)

BRAMBLE SHAFT
Mine
ST 4410 5814
L 1m VR 1m Alt 112m

Situated on the edge of the common, the top of a ginged shaft is visible among the brambles.

PALLET SHAFT
Lost Mine
ST 4421 5806
Alt 122m

In December 2002 a blocked shaft, covered with a wooden pallet was noted in a shallow gruff. The shaft was not descended and the gruff now appears to have been largely filled in, although possible confusion with DGW Mine cannot be entirely ruled out..

Ref: P. G. Hendy, Star Mine and Related Sites Part 1, WCC Jnl 27 (289) pp 143-145 (2004)

OLD SHAFT 2
Mine
ST 4426 5808
VR 10m+ Alt 123m

This shaft lies in the corner of the field, 25m north of the remains of an old mining counterweight still visible in 2016 on the other side of the field boundary (the side containing the football pitch). There is no record of descent, although dropped stones can be heard falling for some distance. This is probably the earth-lined shaft christened '**Gaping Ghyll**' by the Sidcot Schoolboys, who just for once deemed it too dangerous to descend.

Ref: P. G. Hendy, Star Mine and Related Sites Part 1, WCC Jnl 27 (289) pp 143-145 (2004) & SSSS Logbook 3 p 65 (1946)

CEMETERY SHAFT
Mine
ST 4436 5809
Alt 55m

Located 100m south of the cemetery and clearly marked on old OS Maps as 'Old Shaft', this may be the site of a lost Cornish shaft.

Ref: P. Burr, Mines and Minerals of the Mendip Hills, Volume 1, MCRA p 311 (2015)

SHIPHAM

Shipham miners literally sank shafts everywhere and collapses have been recorded under driveways, roads, pavements and gardens, and even, on rare occasions, inside houses. Some have been quite deep, and one even temporarily trapped a Rover 2500. The dense clustering of the shafts, is an unfortunate legacy of a haphazard period of intense mining, when shafts were sunk by privateers exploiting tiny plots of land. Most were filled in, or covered over, in just four days during 1911-12, which probably explains the frequency and number of collapses. The local council was once called upon to fill in three separate shafts situated within a mere 10m of each other, and many more shafts were uncovered and filled during construction of a pipeline through the village. In fact, so many collapses have been reported that there are simply far too many to record individually, so only the most noteworthy mines are mentioned here. Further information on more than eighty others is available from the MCRA website (www.mcra.org).

ROWBERROW BEACH

The term 'beach' is a corruption of batch, the local term for a mining spoil tip and the fields immediately north of the village, and to the east of the cemetery, are collectively known as Rowberrow Beach. Numerous collapses have occurred in these fields, and aerial photos taken before the land was improved, show a veritable moonscape littered with literally hundreds of shafts. One small shaft, **Rowberrow Beach Shaft** (ST 4450 5820, VR 3m), was dug by ACG in 1971, but it yielded only a choked rift. The shafts (known locally as gruffs), rarely exceeded

10m in depth, and most are either capped with flagstones, or have been filled in. Burr has suggested that one easterly site, **Rowberrow Lane Mine** (ST 4486 5771), may be the site of a Cornish shaft. This was reputed to be 91m deep and employed an old locomotive to haul the men and ore.

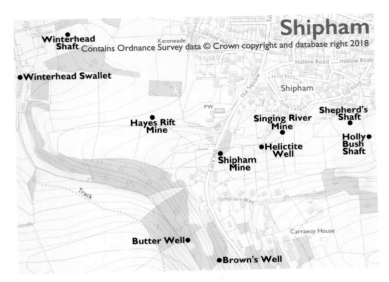

WINTERHEAD

WINTERHEAD HILL SWALLET
ST 4348 5718

Sink
Alt 105m

A stream sinks beneath a hedge in marshy ground.

Ref: C & C Dockrell, Additions to the Complete Caves of Mendip, ACG N/L pp 1-2 (Jun 1992)

WINTERHEAD SWALLET
ST 4383 5752

Sink / Mine
Alt 95m

An infilled collapse, located in the stream bed adjacent to Winterhead Farm. At one stage, pit props were apparently visible, so it was presumably an old mine working. The stream is fed by a series of intermittent springs which rise on the east side of Winterhead Hill. Some were deemed worthy of a name, including the healing well of **Butter Well** (ST 4423 5709), and **Brown's Well** (ST 4428 5706), both of which appear on old maps and are mentioned in Knight's Heart of Mendip.

Ref: C & C Dockrell, Additions to the Complete Caves of Mendip, ACG N/L pp 1-2 (Jun 1992)

WINTERHEAD SHAFT
Mine
ST 4393 5763
L 90m VR 27m Alt 102m

Although originally a much older working, the mine was substantially enlarged by Cornish mining engineers c.1869-1870, one of whom apparently taught a very hands-on version of chemistry at Sidcot School! It consists of an impressive 26m deep Cornish shaft (which ends in a boulder choke) and two horizontal galleries at the 12m level, both of which intercept the earlier workings. It is sometimes called Pendarves Shaft, following a visit by Sir Arthur Pendarves Vivian, KCB, a prominent member of the mine-owning Vivian dynasty, who had concerns in Swansea and Cornwall. The shaft was excavated by ACG down to the galleries level during 1985 and 1986, and the lower section of the shaft was uncovered in 1990.

Ref: Mendip Underground, MCRA pp 450-451 (2013)

WINTERHEAD MINE
Mine
ST 4390 5764
L 30m VR 14m Alt 103m

First noted in 1984, following a surface collapse, this mine was explored by ACG who excavated a circular ginged entrance shaft to gain a roomy east-west rift, with a blocked shaft and passage below. The entrance was subsequently permanently capped and can no longer be positively identified.

Ref: C. Dockrell, Winterhead Mine, ACG N/L 48 pp 6-8 (1985/86)

SHIPHAM VILLAGE & GLOVERS FIELD

HAYES RIFT MINE
Mine
ST 4415 5742
L 100m VR 25m Alt 120m

Located in a field with plenty of evidence of mining activity, this shaft gives access to a number of galleries worked for calamine and lead. Half-inch diameter shot holes are visible throughout the mine, which suggest a 19th century date, although its origins may have been much earlier. The mine was explored during 2010 following the collapse of the capping stone.

Ref: Mendip Underground, MCRA p 167 (2013)

SHIPHAM MINE
Tunnel
ST 4431 5730
L 25m VR 3m Alt 152m

Now blocked, the entrance to this 'mine', which Stanton described as *'a queer place altogether'*, is located among bushes some 20m from the main road. It was excavated through Red Marl during the late 17th century, apparently in an ambitious and clandestine attempt to intercept the water source in Helictite Well. It was visited by both SSSS and ACG who described a tight entrance and short step into an eastward-trending horizontal tunnel carrying a small stream. SSSS traced the water to a small stagnant pool located 80m to the north-west and 17m below.

Ref: W. I. Stanton, Logbook 4 p 82 (1948) & C. Richards, Two Former Water Supplies in Shipham Village, ACG N/L pp 6-9 (Jan 1972)

HELICTITE WELL
Well
ST 4441 5732   L 26m VR 11m Alt 150m

Located on the north side of Cuck Hill and also recorded as Helictite Mine, Womble Well and Womble Mine, a manhole-covered, 9m deep circular walled shaft yields two superimposed galleries cut through Old Red Sandstone. There are strange formations in the Lower Gallery resembling helictites and there is a lead pipe in situ, which may once have carried water to one of the houses in the square. Two fields further east, and 30m higher, is a walled spring which may be the un-failing one recorded by Knight in his Heart of Mendip. The water apparently once ran across the fields to the village, but was later captured into a mine – possibly entering Singing River Mine at the Great Hall.

Ref: C. Richards, Two Former Water Supplies in Shipham Village, ACG N/L pp 6-9 (Jan 1972) & M. Ireland, Helictite Well, Shipham, BEC Belfry Bull 53 (518) pp 25-28 (2004)

SINGING RIVER MINE
Mine
ST 4447 5736   L 1000m VR 28m Alt 152m

Also known as Shipham Old Mine Well and Water Mine, this intricate and interesting mine is one of the longest still open on Mendip and offers the visitor a veritable maze of passages, chambers and rifts to explore. Rediscovered by ACG in 1971, the main passages are laid out along an east-west axis and the entrance shaft enters the mine roughly at the centre. To the east lies a series of impressive and watery chambers, while to the west, an active streamway can be pursued for almost 100m, crossing a series of attractive deep blue pools. The mine was originally worked in the 18th century and continued until 1894. One Cornish shaft is said to have reached a depth of 55m and was serviced by a steam engine. Mid-19th century workings were revealed here below the flooded level in 1916-18 when the shaft was partially cleared during extensive engineering works which modified the Great Hall and Stinking Gulf for use as a reservoir. These plans were abandoned when water levels were found to be too low. Shipham dowsers claimed to have traced the course of the Banwell Spring back to a spot in Longbottom Valley near Shipham, where, in 1813, a shaft was sunk in search of coal. However all attempts to prove a connection with the area (including the stream seen in Singing River Mine), have so far failed.

Ref: Mendip Underground, MCRA pp 320-325 (2013)

HOLLY BUSH SHAFT
Mine
ST 4469 5734
L 170m VR 25m Alt 163m

A 20m deep shaft enters an extensive series of 18th and 19th century calamine workings which connect with Shepherd's Shaft, 30m to the north-west. The passages display plenty of evidence of mining techniques including shot holes, pick marks and blackening from the use of tallow candles. First noted by the ACG in 1971, it was entered by BEC diggers in 2003.

Ref: Mendip Underground, MCRA pp 174-175 (2013)

SHEPHERD'S SHAFT
Mine
ST 4464 5738
L 31m VR 11m Alt 158m

A 15m shaft, which connects with the western gallery of Holly Bush Shaft. Currently (2013) sealed by large concrete slabs, it was connected to Holly Bush Shaft in 2004.

Ref: M. Ireland, Holly Bush Shaft, Shipham – Recent Explorations, BEC Belfry Bull 53 (518) pp 18-24 (2004)

COMPTON BISHOP & CROSS

The village of Compton Bishop is overlooked by the notable landmark of Crook Peak to the west, and the long whaleback ridge of Wavering Down to the east. The hills fall within the bounds of the Crook Peak to Shute Shelve SSSI, a 332 hectare area of geological and biological importance. A number of small but interesting caves are located along the south-eastern approach to Crook Peak, several of which have yielded important archaeological remains. Some of the caves are lightly dolomitized and may have been formed by hydrothermal solution. The parish also includes the hamlet of Rackley (formerly Radeclive or Redcliff), a port founded by papal bulls and Royal Charters in the 12th century. This was once an important trading post on the River Axe and may even have been the point from which the Romans exported the lead they extracted from Charterhouse. The hamlet now lies well to the north of the diverted river, and almost no sign of the port now remains.

DUNNETT SPRINGS
Spring
ST 4009 5471
Alt 6m

Also known as Dunyeat Springs, several small springs which rise in Dolomitic Conglomerate in a copse and neighbouring small field, are piped to a central tank. The overflow debouches into the River Yeo, along with water from another overflow which enters the river 30m upstream. The water is reputed to be extremely hard.

Ref: N. Barrington, W. Stanton, The Complete Caves and a view of the hills, p 63 (1977)

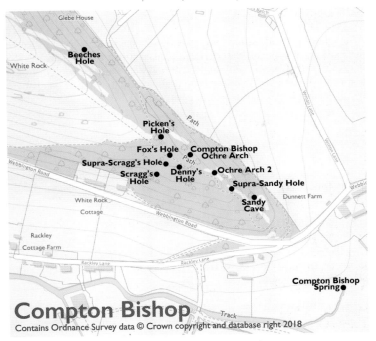

COMPTON BISHOP SPRING
ST 3989 5479

Spring
Alt 7m

Said to run strongly for nine months of the year, in wet weather this spring emerges as a forceful bubble below a low cliff.

Ref: N. Barrington, W. Stanton, The Complete Caves and a view of the hills, p 63 (1977)

SANDY CAVE
ST 3976 5493

Cave
L 15m VR 3m Alt 32m

A cave with umpteen names including, Sandy Hole, Fox Hole, Sandy, The Devil's Hole, Fox's Hole (Compton Bishop), Devil's Cave, Crooks Peak Cave, Foxe's Gulley and Foxes Hole (Compton Bishop). The cave is located in an old quarry situated below, and to the north of the main track leading up to Denny's Hole. First dug by WNHAS in 1903, it consists of a large entrance leading to a series of small, sandy chambers containing deposits of yellow ochre and chalcedony. There is a low choked archway in the quarry face directly opposite.

Ref: A. Legg, The caves and sites of spelaeological interest around Compton Martin [Bishop], MNRC N/L 38 pp 8-12 & N/L 39 pp 5-9 (1992)

SUPRA-SANDY HOLE Cave
ST 3974 5494   L 10m VR 3m Alt 40m

Also known as Supra-Sandy Cave, a low, wide shelter in a cliff just to the north of the main summit path yields a short crawl. It was dug briefly by SSSS in 1945, who concluded that it would connect with a small depression located a few metres away on the ridge. The crawl passes directly over Sandy Cave where tapping sounds can clearly be heard.

Ref: A. Legg, The caves and sites of spelaeological interest around Compton Martin [Bishop], MNRC N/L 38 pp 8-12 & N/L 39 pp 5-9 (1992)

COMPTON BISHOP OCHRE ARCH 2 Cave
ST 3971 5496   L 2m VR 1m Alt 44m

This low, horseshoe-shaped choked archway is located among cliffs to the north of the main path and a short distance south-east of Compton Bishop Ochre Arch.

COMPTON BISHOP OCHRE ARCH Cave
ST 3969 5499   L 2m VR 1m Alt 42m

Located amongst crags north of the path and to the south-east of Picken's Hole, this low, wide steeply-inclined phreatic alcove is heavily choked with ochreous fill spilling from a badger hole above.

DENNY'S HOLE Cave
ST 3967 5497   L 91m VR 13m Alt 48m

This cave is the large obvious opening located immediately south of the main ridge path. A large entrance leads down into a roomy chamber (5m high and 10m in diameter) from where several small passages lead off. The cave was first described by Alexander Catcott in 1758 and in the 19th century it was sometimes referred to as Phelps' Hole, after another antiquarian. It was used as a hideout by a covert Auxiliary Unit of the Home Guard during WWII and was the scene of largely fruitless SSSS digging in the 1940s. At one time it was thought that the lack of archaeological material suggested that the cave may have been breached only recently. However digging by persons unknown in the 1970s uncovered a layer of bones and charcoal (from which a flint was later recovered), and it may be that other material was looted by locals who Catcott termed '*adventurous rustics*'. He personally referred to the cave as Dennis's Hole, believing it to be derived from a nearby, but long-vanished shrine to St Dennis. Denny is simply a corruption. Recent work has suggested a hypogene origin for this and other nearby caves, formed by the mixing of deep thermal waters with oxygenated surface water.

Ref: Mendip Underground, MCRA p 110 (2013) & S.A. F. McArdle, P. L. Smart, Geomorphology of Denny's Hole and associated caves, Crook Peak West Mendip: A newly recognised hypogene cave complex, UBSS Proc 28(1) pp 65-102 (2019)

FOX'S HOLE
Cave
ST 3966 5498　　　　　　　　　　　　　　　　　　L 18m VR 3m Alt 50m

Also recorded as Foxes Hole, Hole X and The Slitter, this tight vertical fissure is actually located directly on the footpath itself. Although it yields only a low chamber, the presence of vapour, seen rising during cold weather, suggests a blocked connection with nearby Denny's Hole. Shotholes in the entrance prove that it was artificially widened.

Ref: A. Legg, The caves and sites of speleological interest around Compton Martin [Bishop], MNRC N/L 38 pp 8-12 & N/L 39 pp 5-9 (1992)

SUPRA-SCRAGG'S HOLE
Cave
ST 3966 5497　　　　　　　　　　　　　　　　　　L 2m VR 2m Alt 48m

Also known as Southall's Hole, Cindy's Cave and A Hole, this cave is situated in thick scrub a few metres south-west of Denny's Hole. The entrance is a low wide slot, partially concealed by detached boulders and yields a short rising crawl. The cave was apparently excavated by SSSS in 1930, who allegedly found the remains of six human skeletons, said to have been examined by Dr N. Cooper of Winscombe. Unfortunately evidence to substantiate this claim has yet to materialise and it may have been confused with another, altogether different site.

Ref: A. Legg, The caves and sites of speleological interest around Compton Martin [Bishop], MNRC N/L 38 pp 8-12 & N/L 39 pp 5-9 (1992)

SCRAGG'S HOLE
Cave
ST 3964 5496　　　　　　　　　　　　　　　　　　L 12m VR 3m Alt 43m

This roomy tunnel, which is located in dense scrub to the south of the path and 30m south-west of Denny's Hole, was excavated by SSSS in 1943-45, 1948-49 and in 1953, and also by the ACG in 1951, and again in 1961-62. Evidence of Romano-British occupation was uncovered, but there is no evidence of anything pre-dating this period, which suggests that the cave was unknown to earlier inhabitants and therefore probably only collapsed open relatively recently. The name was provided by a local cottager, although its precise origin remains unknown.

Ref: Various Sidcot School logbooks including a separate logbook entitled 'Scragg's Hole and Bos Swallet' available to view online via the MCRA at www.mcra.org.uk/logbooks

Crook Peak and Cross

Contains Ordnance Survey data © Crown copyright and database right 2018

West Mendip · Compton Bishop & Cross

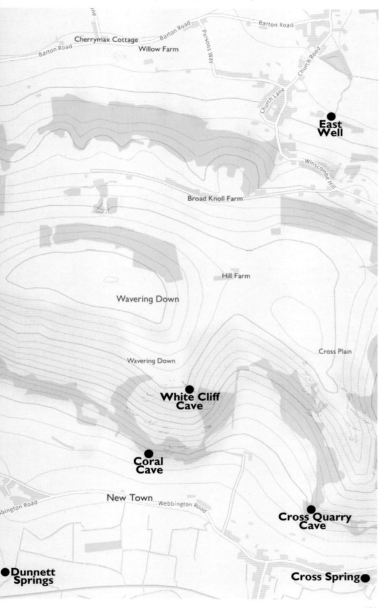

131

PICKEN'S HOLE   Cave
ST 3965 5502   L 6m VR 1m Alt 50m

Located in woodland to the north of the path, this cave is a Scheduled Ancient Monument and is probably the most carefully excavated hyena den and assemblage site in Britain. There is an abundance of spotted hyaena bones, and a high degree of chewing on the majority of other recovered bones. Although small, the cave is unusual in that it contains two large bone assemblages, brought to the site at different periods of the Late Pleistocene and is of considerable importance owing to its clear, well-stratified sequence of deposits and faunas, all dating from the Devensian. The deposits are spread over six identified layers, which have helped to shed considerable light on the prevailing climatic conditions. It was excavated by UBSS in 1961 to 1967, and is named after Mr. M. J. Picken, who discovered it while studying badger setts in the area. Badgers still visit the cave today. Among the most important finds were Pleistocene animal remains, plus human teeth and Mousterian flint flakes, including two hammerstones. A Neolithic human premolar has been dated to c.4800 BP, while the rich Layer 3 fauna was radiocarbon dated to 34,265 (+2600/–1950) years BP. The finds extracted (spotted hyena, lion, arctic fox, woolly mammoth, woolly rhinoceros, horse, reindeer, suslik and tundra vole) are consistent with a cold, tundra climate. The lowest layer produced finds of reindeer, red deer and bear, consistent with a cool forest environment c.40,000 years ago. Reports that the upper layers contained Roman-British pottery are incorrect, the cave having been confused with excavations at nearby Scragg's Hole. The confusion apparently arose when a local farmer (incorrectly) suggested Scragg's Hole as an alternative name for Picken's Hole.

Ref: E. K. Tratman, Picken's Hole, Crook Peak, Somerset, A Pleistocene Site, UBSS Proc 10(2) pp 112-115 (1964), G. J. Mullan, Radiometric Dating of Samples from Picken's Hole, UBSS Proc 27(3) pp 261-265 (2018), K. Scott, The Large Vertebrates from Picken's Hole, Somerset, UBSS Proc 27(3) pp 267-313 (2018), R. M. Wragg Sykes, Picken's Hole, Crook Peak, Somerset: A Description of the Lithic Collection and its Probable Late Middle Palaeolithic Context, UBSS Proc 27(3) pp 315-338 (2018) & A. M. ApSimon, G. J. Mullan, The Human Teeth from Picken's Hole, UBSS Proc 27(3) pp 339-341 (2018)

BEECHES HOLE   Cave
ST 3954 5514   L 3m VR 1m Alt 58m

This low, choked phreatic passage is located in woodland on the northern slopes of the ridge. A likely untouched hyaena den, it has clear archaeological potential and should be treated accordingly.

Ref: N. Barrington, W. Stanton, The Complete Caves and a view of the hills, p 38 (1977)

WEBBINGTON QUARRY CAVE   Cave
ST 3896 5521   Alt 50m

Also recorded as Compton Bishop Quarry Cave, this tiny hole was located in the west face of a small abandoned quarry. It was lost in 1968 when the quarry was partially filled in and there is no visible sign of it today (2015).

CROOK PEAK RIFT
Fissure
ST 3888 5567   L 2m VR 2m Alt 163m

This long, choked rift, formed in sandy decomposing dolomitic limestone, lies adjacent to the footpath and 200m south-east of the summit of Crook Peak. Allegedly first probed in 1966, it was dug again c.1971. Largely backfilled, all that can be seen today are a line of small, bramble-filled pits, barely distinguishable from the surrounding ground. The 1966 dig may actually have taken place at **Crook Peak Hole** (ST 3880 5579, L 2m VR 1m), a low inclined slot hidden behind brambles in the south-eastern corner of the low cliff which forms the distinctive summit.

Ref: W. I. Stanton, Logbook 11 p 202 (1969) & N. Barrington, W. Stanton, The Complete Caves and a view of the hills, p 58 (1977)

WOLF DEN
Cave
ST 3955 5634   L 4m VR 3m Alt 135m

Also known as the Wolf's Den, Barton Cave and Wringstone Rocks Cave, this cave has two superimposed entrances which unite in a small chamber. It lies in a narrow gully, which due to dense undergrowth is best accessed from above. It was excavated by Stanton in 1944-45, who unearthed horse and wolf bones dating to the Late Pleistocene, which are now visible in Wells Museum. 20m to the west, there is another, slightly larger gully containing a small overhanging shelter with a blocked crawl beneath (**Upper Barton Shelter**, ST 3953 5634). A line of small ginged pits situated on the plateau immediately above, are the remains of **Wringstone Rocks Lead Mine** (ST 3953 5632, L 1m VR 1m).

Ref: N. Barrington, W. Stanton, The Complete Caves and a view of the hills, p 177 (1977) & N. Richards, Personal communication

BARTON SHELTER
Cave
ST 3952 5635   L 8m VR 3m Alt 130m

Also recorded as Badger Hole and located 50m north-west of Wolf Den and at a slightly lower level, this roomy shelter contains three small phreatic tubes. Only the central tube is penetrable, which closes down after a short hands and knees crawl. It was recorded by Stanton in 1945 and excavated to its current length by ACG c.1950.

Ref: N. Barrington, W. Stanton, The Complete Caves and a view of the hills, p 36 (1977)

BARTON DROVE SHAFT
Mine
ST 4063 5634   L 2m VR 2m Alt 140m

Little is known of this isolated 'shaft', which doesn't appear on OS maps until the 1920s. Recorded as 2m deep, and 2m in diameter, it is thought to have been sunk in search of water and has now been filled in.

CROSS PLAIN FISSURE
ST 4150 5570

Mine / Cave
Alt 145m

In 1990, several substantial pieces of ribbed flowstone were identified in the spoil of some shallow, long-abandoned mine workings. Although no official record of mining exists, they were probably small-scale ochre workings, which presumably followed, or intersected, a natural cave passage. The area has become increasingly overgrown in recent years and the exact site can no longer be identified with any certainty.

Ref: T. R. Shaw, A once open fissure on Cross Plain, Wavering Down, WCC Jnl 20 (224) p 134 (1990)

WHITE CLIFF CAVE
ST 4088 5550

Cave
L 8m VR 2m Alt 140m

Located midway along the prominent cliff situated at the top (western) edge of Bourton Combe, this short phreatic through trip is best approached from above. It was first recorded by WCC in 1971 and excavated by ACG in 1994, who found a coin farthing dated 1694, along with miscellaneous bones and pottery. A sloping descent leads to a hands and knees crawl which exits via an artificially widened window in the cliff face. A third entrance, located directly below, is currently blocked. The finds are in Axbridge Museum.

Ref: N. Barrington, W. Stanton, The Complete Caves and a view of the hills, p 173 (1977), W. I. Stanton, Logbook 12 p 69 (1971) & M. J. Norton, White Cliff Cave, ACG N/L p 22 (Summer/Autumn 1995)

CORAL CAVE
ST 4069 5524

Cave
L 92m VR 25m Alt 47m

Located in woodland at the base of a low outcrop, a locked manhole yields a 15m pitch into a large chamber, where a roomy passage can be followed for 30m to a mud choke. In very wet weather, the bottom of the main chamber floods to a depth of 3m forming a lake 6m or more across. The somewhat misleading name was derived from the splash deposits of stalagmite, giving the appearance of branched coral, found at the lowest point of the cave. The cave was opened c.1905 by a farmer obtaining stone for a drinking place for his cattle. Originally entered by an adventurous local, it was explored by a team led by Baker and Balch and closed briefly following an accident involving Tony Richardson, a Sidcot schoolboy in 1945, who let go of the rope and plummeted down the entrance shaft. Fortunately, despite having initially been knocked unconscious, Richardson was recovered from the shaft having suffered nothing more than a broken rib. Stanton on the other hand, the leader of this 'unofficial trip', was sent home from school for a week and temporarily suspended from all caving activities. During an eventful year he was later reproved for drinking cider on V.E Day! The terminal mud choke was dug in vain by ACG and SSSS between 1961 and 1965. Gerard Platten and his friends once found a dead sheep propped up inside the cave wearing a boiler suit and the cave was later famously owned by the comedian Frankie Howerd, who lived in Wavering Down House until his death in 1992. Titter ye not!

Ref: Mendip Underground, MCRA p 101(2013)

CROSS QUARRY GEODE
Cave
ST 4137 5497
L 2m VR 1m Alt 60m

Reached by an easy scramble, this cave is the sole surviving penetrable cavity in the quarry. It is basically a geode, lined with large calcite crystals, and would make an excellent hiding place. Judging from old photographs it appears to sit directly above the spot once occupied by Cross Quarry Cave.

Ref: N. Barrington, W. Stanton, The Complete Caves and a view of the hills, p 58 (1977)

CROSS QUARRY CAVE
Cave
ST 4137 5495
Alt 50m

Located at quarry floor level, directly below Cross Quarry Geode, this small, dripstone-lined phreatic cavity was discovered and quarried away c.1925. In 1969, Stanton recorded several waterworn, ochreous cracks and tubes in the southern part of the east face of the quarry. However the quarry appears to have been partially infilled since and these are no longer clearly visible.

Ref: N. Barrington, W. Stanton, The Complete Caves and a view of the hills, p 58 (1977) & W. I. Stanton, Logbook 11 p 205 (1969)

CROSS SPRING
Springs
ST 4159 5469
L 7m VR 7m Alt 9m

Also known as Cross Well and South Marsh Water Works, this capped, 7m deep well in Dolomitic Conglomerate, was sunk on the site of several large natural springs in 1898. Water from other nearby springs is piped to it, including one at Rackley. A gold Bronze Age bracelet, made from twisted wire, was discovered here during construction of the water works, which has been taken as evidence of ritual deposition. It is now on display in the British Museum. There are several other wells and springs in the village, one of which (**Old Forge Well**, ST 4128 5479), was once a renowned healing well. Said to contain water so '*exceptionally pure*' that it was sought after by doctors in neighbouring Axbridge, it was reputedly located in the garden of the old forge (now Fairfield).

Ref: N. Barrington, W. Stanton, The Complete Caves and a view of the hills, p 58 (1977) & W. I. Stanton, Logbook 11 p 205 (1969)

AXBRIDGE

Axbridge is a very old borough and the name probably derives from Axanbrcyg, a 9th century name, meaning a bridge over the River Axe. Occupation dates back to the earliest times, as evidenced by the many flint tools which have been found on the surrounding hills, but the earliest mention of Axbridge itself dates from AD 911, when the town is described as possessing a fortified wall over 500 yards in length. Royal borough status enabled the town to send members to parliament and it even had its own mint, with coins showing the town's symbol (the Lamb and Flag), an emblem later made famous by the ill-fated Duke of Monmouth. The most famous building is King John's Hunting Lodge, a wool merchant's house dating to around 1460, which once comprised shops, living areas and workshops. It is owned by the National Trust who lease it to the Axbridge and District Museum Trust, a group of dedicated local volunteers who operate it as a local museum. It includes exhibits relating to the local geology and history. References to ochre mining date back to at least 1685, and mining continued right through until the 1930s. Several important caves were intercepted, including the famous 'Lost Cave of Axbridge', which contained a chamber reputedly '*bigger than Axbridge Square*'. This 'lost' cave was finally rediscovered in 2011. All the sites lie to north of the village, where the hills can be sub-divided into four distinct sections, each separated from one another by narrow defiles. Shute Shelve Hill lies within the Crook Peak to Shute Shelve Hill SSSI, a 332 hectare location of geological and biological interest. Axbridge Hill and Fry's Hill, form another 64 hectare SSSI, as does Cheddar Wood, an 87 hectare site managed by the Somerset Wildlife Trust. The latter is the largest surviving remnant of King Edmund's 10th century hunting forest.

SHUTE SHELVE HILL

Shute Shelve Hill was extensively mined for ochre during the early 1920s. Collectively known as the Rose Wood mines, these were the last of several workings developed by Richard Myatt, and were equipped with narrow gauge railways and a 125m long cableway with buckets designed by John Weare, the local undertaker!

CARCASS CAVE
ST 4239 5530
L 150m VR 46m Alt 115m

Cave

Instantly recognisable by the large wall of digging spoil, an in situ monorail, and the fenced-off shaft, this complex little cave has three entrances and consists of a series of chambers and shafts aligned along a north-south fault. There is a large geode, a beautiful gour pool and some excellent dog-tooth crystals to admire. The 'Swiss cheese' like structure is explained by the ancient hydrology of the area. Under phreatic conditions, water flowing under high pressure through nearby Shute Shelve Cavern was forced upwards through cracks and fissures to create Carcass Cave. The shaft entrance, the only part of the cave obviously dug by the ochre miners, was re-excavated by ACG between 1996 and 1998, and a sustained digging effort by ACG, BDCC and MCG between 2001 and 2005 opened up the rest of the cave.

Ref: Mendip Underground, MCRA p 81 (2013)

SHUTE SHELVE CAVERN
Cave
ST 4239 5535   L 249m VR 56m Alt 106m

Also known as Axbridge New Cave, this geologically interesting cave contains some of the largest chambers to be found in the west Mendip area. Although there is clear evidence of ochre mining in the entrance passages, the main chambers are entirely natural and probably represent a section of a large phreatic loop which drained water from a catchment in the Lox Yeo valley to the north, to an ancient resurgence near Axbridge. Nearby Carcass Cave is another fragment of this drainage system. A low steep entrance passage leads to a drop through boulders into the 30m long Flat Room, where fine dog-tooth crystals can be viewed. Another climb down reaches Box Tunnel, 40m long, 12m wide and up to 10m high. The passages beyond quickly end in digs. The cave was first entered by local ochre miners in the 1920s, but was subsequently lost. It was rediscovered by ACG in February 1992 and the size of the newly discovered chambers prompted speculation that it might be the famous 'lost cave' which had been described by the miners in the late 1940s as being *bigger than Axbridge Square*. The subsequent discovery of Axbridge Hill Cavern proved otherwise.

Ref: Mendip Underground, MCRA pp 315-317 (2013)

ROSE WOOD BOUNDARY CAVE
Cave
ST 4242 5536   L 7m VR 4m Alt 112m

Located at the foot of a small cliff (the remains of a quarried cavern), a steep drop gives access to a small natural chamber backfilled with miners' infill to a depth of at least 4m. There is a second penetrable hole (**Upper Rose Wood Boundary Cave**, L 5m VR 2m Alt 115m), located near the top of the cliff, close to the fence on the left-hand side, and a well-preserved miners' hut on the flat ground immediately below. There are numerous other small pits and trial trenches in the immediate vicinity, some of which display the hallmark square and rectangular cut associated with Richard Myatt.

Ref: M. Clarke, N. Gregory and A. Gray, Earth Colours – Mendip and Bristol Ochre Mining, MCRA, p 245 (2012)

Shute Shelve Cavern. Photo by Steve Sharp

DICK TURPIN'S CAVE   Lost Cave

A fabled 'lost' cave said to exist on Shute Shelve Hill, and visited by a friend of John Chapman's father, named Faulkner, who remembered playing inside it c.1900. Its exact location is unknown. Interestingly there are several other lost 'Dick Turpin's Caves' dotted around the British countryside, the most famous being in Epping Forest, close to where the real Dick Turpin practised his nefarious art. The Shute Shelve variant is probably just a childish name for one of the vicinity's better known caves (possibly Shute Shelve Cavern).

Ref: D. J. Irwin, The Lost Caves of Mendip, BEC Belfry Bull 505 pp 31-46 (1999)

MYATT'S MINE   Mine
ST 4270 5549   L 23m VR 10m Alt 190m

Perhaps better known as The Rift, this is a 45m long, 6m wide cutting, developed along a natural fissure. The cutting shows scalloping in several places while the cave exhibits waterworn rock surfaces throughout. The presence of goethite (yellow ochre) stalactites in a small muddy cave at the far end indicates that the ochre was deposited inside the caves through fissures from above. Renamed in honour of Richard Myatt, the mine was served by a 50m long narrow-gauge railway.

Ref: M. Clarke, N. Gregory and A. Gray, Earth Colours – Mendip and Bristol Ochre Mining, MCRA, pp 234-235 (2012)

SHUTE SHELVE QUARRY HOLE   Mine
ST 4268 5546   L 2m VR 1m Alt 190m

Similar in many ways to Myatt's Mine, this nearby overgrown quarry displays a number of waterworn features, including a significant phreatic pocket which has clearly been worked for ochre.

FENNEL HOLE   Cave
ST 4247 5476   L 6m VR 2m Alt 43m

Located in woodland, 60m south-east of the large picnic lay-by, this cave consists of a small chamber containing several large calcite crystals.

Ref: N. Barrington, W. Stanton, The Complete Caves and a view of the hills, p 77 (1977)

BYPASS GEODE   Cave
ST 4251 5478   L 5m VR 2m Alt 52m

Also known as Axbridge Bypass Cave, this small cavity, lined with large calcite crystals, was uncovered during road works in 1966. Clearly visible for many years it now lies almost completely hidden in undergrowth behind a high fence. There have been several rock falls in recent years and the entrance appears to have been buried.

Ref: N. Barrington, W. Stanton, The Complete Caves and a view of the hills, p 48 (1977)

AXBRIDGE HILL

Axbridge Hill is home to two major groups of caves. The first, which plays host to the fabled Lost Cave of Axbridge, can be found situated among a group of ochre pits at the western end of Axbridge Hill, at the point where the tramway leading up from Axbridge Station terminates. The second is home to probably the most important mine on the hill and lies in woodland above 'Hillside', close to the narrow defile separating Axbridge Hill from Fry's Hill. A drove running through this defile can be followed up onto Fry's Hill, then onward to meet with Callow Drove and routes down to Winscombe.

THE LOST CAVE OF AXBRIDGE

Possibly discovered as long ago as the 17th century, this famous 'lost cave' was described by the miners in the late 1940s as being *'bigger than Axbridge Square'*. It was said to contain a large underground lake. The search for this fabled cave began in the 1950s, when ACG discovered Triple-H Cave and Large Chamber Cave. In 1992, on an adjacent hill, ACG broke into Shute Shelve Cavern and for a while its large chambers were thought to represent the lost cave. Then a clue to its real location was revealed inside the late Willie Stanton's Log Books. In 1965 he recorded that he was informed by Tim Chard (the landowner's son), that the 'big chamber' lay close to Large Chamber Cave and in front of several small holes in a limestone cliff which had been excavated by the miners (Axbridge Hill Ochre Pits). In February 2011 a digger excavated this area but only a small cave was located. Digging by hand continued however and in April 2011 the lost cave was finally rediscovered. The cave was duly christened Axbridge Hill Cavern (see below) and the chamber is in fact about one third of the size of Axbridge Square!

LARGE CHAMBER CAVE   Cave
ST 4289 5493   L 16m VR 5m Alt 111m

Located immediately to the left on entering the ochre pits enclosure, this rather ironically-named cave consists of a short crawl into a small chamber. It was discovered by the ACG in 1954.

Ref: R. J. Weare, Large Chamber Cave, Axbridge Hill Report, ACG Jnl 2 (2) pp 36-38 (1954) & N. Barrington, W. Stanton, The Complete Caves and a view of the hills, p 106 (1977)

AXBRIDGE HILL OCHRE PITS   Cave
ST 4288 5493   L 14m VR 3m Alt 111m

Situated in the fenced-off area excavated by the digger in 2011, these cavities, which were excavated in the 1930s, are the *'several small holes'* recorded by Stanton in 1965. The largest is 4m long and 3m deep and along with Axbridge Hill Cavern contain probably the best remaining ochre deposits seen on Mendip.

Ref: W. I. Stanton, Logbook 11 p 63 (1965) & M. Clarke, N. Gregory and A. Gray, Earth Colours – Mendip and Bristol Ochre Mining, MCRA, pp 236-238 (2012)

Axbridge Hill Cavern. Photo by Steve Sharp

Somerset Underground · Volume 2

Axbridge

Contains Ordnance Survey data © Crown copyright and database right 2018

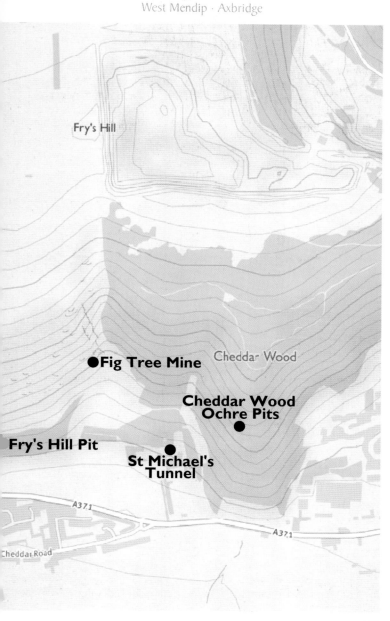

AXBRIDGE HILL CAVERN
Cave
ST 4288 5493 L 62m VR 14m Alt 111m

This gated cave is the fabled 'Lost Cave of Axbridge', much spoken of by ochre miners in the area in the last century. A sloping 2m crawl over the top of miners' deads, gives access to the Main Chamber, 22m long, 8m wide and up to 8.5m high. There are excellent ochre deposits throughout, while a broad dry hole in the centre of the chamber probably once held the underground lake described by the early explorers. This has long since vanished. Rediscovered in 2011, the cave revealed a wealth of evidence from the ochre mining industry, including an old shovel, pit props, a possible candle tin, and pick and candle marks. Best of all was a miner's boot, which was identified by Clarks Shoe Museum in Street as a Derby Boot dating from the late 1880s to c.1920.

Ref: Mendip Underground, MCRA pp 43-44 (2013)

TOAD PIT
Mine
ST 4287 5497 L 10m VR 6m Alt 130m

Situated in dense woodland 40m north of the main ochre pits enclosure, a 6m deep shaft yields a short passage in ochreous fill. This is probably the site christened Kentish Cave, which was dug briefly by Kent-based ACG members, and there is another mined pit nearby with a low, choked hole which may yield results.

TRIPLE-H CAVE
Cave
ST 4294 5494 L 25m VR 6m Alt 115m

Until recently concealed in woodland, this cave now lies within a fenced-off, bramble-filled hollow located close to the tramway on the open hillside. Excavated in 1952 on the basis of experimental echo sounding tests, this was the first of the sites examined by ACG during their search for the fabled Lost Cave of Axbridge. Initially dug to a solid rock floor at a depth of 13m, a subsequent collapse revealed the cave proper, which consists of a roomy shelter with a side passage connecting to a tight second entrance. Known as **The Backdoor**, this no longer appears accessible. An archaeological investigation conducted from 1953-55 revealed Late Pleistocene remains, including bear, hyaena, wolf and rhinoceros.

Ref: R. J. Weare, Triple-H and associated work on Axbridge Hill, ACG Jnl 2 (1) pp 13-14 (1954) & N. Barrington, W. Stanton, The Complete Caves and a view of the hills, p 164 (1977)

TRAMWAY PITS
Mines
ST 4298 5498 Alt 125m

There are several mined depressions located close to the northern loop of the ochre pits tramway. Some contain choked holes which may repay digging.

LETTERBOX CAVE
Cave
ST 4310 5496 L 12m VR 6m Alt 119m

Like Triple-H Cave, the entrance to this small cave lies on the open hillside and has been fenced-off. Below a mass of near-impenetrable brambles, a small descending tunnel, cut through massive crystalline dripstone, yields a small chamber heavily choked with earth and stones. It was probably opened by ochre miners in the 1920s. Stanton noted that flints were found inside, but this was probably confused with nearby Flint Crevice.

Ref: R. J. Weare, Letterbox Cave, Axbridge Hill Report, ACG Jnl 2 (2) pp 36-38 (1954), W. I. Stanton, Logbook 11 p 62 (1965) & N. Barrington, W. Stanton, The Complete Caves and a view of the hills, p 107 (1977)

FLINT CREVICE
Cave
ST 4312 5502 L 3m VR 2m Alt 128m

This natural narrow crevice, which is difficult to find, lies in an area of dense gorse close to the edge of a small cliff. It was excavated in 1954 by ACG who found over twenty worked microlith flints.

Ref: R. F. Everton, Flint Crevice excavation and adjacent surface assemblages, Fry's Hill, Axbridge, Axbridge Archaeology and Local History Society Journal pp 3-5 (1978) & N. Barrington, W. Stanton, The Complete Caves and a view of the hills, p 78 (1977)

AXBRIDGE OCHRE LOWER MINE
Mine
ST 4320 5505 L 6m VR 1m Alt 96m

The lowest of the mines situated close to the narrow defile separating Axbridge Hill from Fry's Hill, this is a 6m long roomy tunnel mainly formed in cemented angular breccia. The right-hand wall however is in solid limestone and contains a calcite-encrusted shothole. Care should be taken as there may be bones present in the gravel.

AXBRIDGE OCHRE MIDDLE MINE
Mine
ST 4318 5507 L 10m VR 1m Alt 111m

Also known as Bear Pit, this cave is located close to the path leading up to Axbridge Ochre Cavern. A flat-out crawl yields a roomy chamber where parts of a fossilised cave bear have been unearthed in the gravel.

Ref: N. Richards, Personal communication

THE TWINS
Mines
ST 4319 5510 L 3m VR 3m Alt 112m

Recorded on old cave registry sheets, these are two 3m deep holes, spaced 3m apart.

Sidcot Schoolboys in Axbridge Ochre Cavern c.1954.
Charles W. Smith collection, courtesy of Sidcot School

AXBRIDGE OCHRE CAVERN   Cave / Mine
ST 4312 5510   L 123m VR 33m Alt 131m

Also known variously as Axbridge Ochre Pit, Callow Ochre Cave, Callow Ochre Pit Cave, Axbridge Ochre Hole, Callow Ochre Mine, Callow Ochre Cavern and Axbridge Ochre Mine, this working consists of a very deep excavated rift, with an upper network of decorated natural passages. Work to remove the ochreous fill was begun by local farmer John Morgan in 1873, although most of the work appears to have been carried out between 1920 and 1930, initially under the control of mines manager Richard Myatt. A narrow gauge railway was installed, along with a windlass and an incline, and a winding wheel made by Dening & Co. of Chard can still be found in situ. The upper network was first entered c.1921, before the fill below was removed, and can now only be reached by a sheer 6m climb.

Ref: Mendip Underground, MCRA pp 44-45 (2013)

BALCONY CAVE   Cave
ST 4312 5510   L 6m VR 3m Alt 135m

Situated on a balcony overlooking Axbridge Ochre Cavern, this cave consists of a roomy tunnel which terminates in a compacted choke of mud and rocks. Scalloping is much in evidence and there is another smaller hole in the cliff nearby.

WINKS WELL   Spring
ST 4323 5516   Alt 130m

Also known as Wee Willy Winkie's Well and recorded as far back as 1620, this fascinating tiny water source is located on a ledge part way up a rock outcrop. It consists of a small, hand-carved basin, above a tufaceous cascade, and investigations by the ACG in the 1960s uncovered a circular well and a beautifully constructed drainage channel of red earthenware. This, apparently, was the water supply for a self-sufficiency project built by George Cumberland, traces of which have long since vanished. Cumberland was a notable English art collector, poet and lifelong supporter of William Blake, with whom he shared an interest in printmaking. He was also a noted fossil collector, and in 1826 published his *Reliquiae conservatae*, a study of fossil marine organisms.

Ref: A. Everton & V. Russet, Winks Well - Axbridge Archaeology and Local History Society Newsletter 101 pp 10-13 (1987)

FRY'S HILL SHAFTS EAST   Mine
ST 4340 5550   Alt 220m

One of the largest mines in the parish, Fry's Hill Mine was developed by the Mendip Oxide & Ochre Co. during the 1930s. The mine had two entrance shafts (Fry's Hill Shafts East & West, which were also known as Plantation East & West Mine), and a series of four ladders were required to reach the lowest level. The infilled eastern entrance lies among a group of gorse-shrouded ochre pits, located 200m south-east of Fry's Hill Shafts West. A large dump and a barrow run remain visible.

Ref: M. Clarke, N. Gregory and A. Gray, Earth Colours – Mendip and Bristol Ochre Mining, MCRA, p 141 (2012)

FRY'S HILL SHAFT CENTRAL   Mine
ST 4331 5556   Alt 225m

An infilled ochre pit situated midway between Fry's Hill Shafts East and West.

FRY'S HILL SHAFTS WEST   Mine
ST 4329 5564   Alt 230m

The western entrance to Fry's Hill Mine lies among a group of overgrown ochre pits in the north-west corner of the field. Trial shafts were also sunk 300m to the north-east during this period. A depression containing a small pool and spoil heap are all that remains of **Forty Acres Mine** (ST 4353 5579). No further details are known.

Ref: M. Clarke, N. Gregory and A. Gray, Earth Colours – Mendip and Bristol Ochre Mining, MCRA, p 141 (2012)

CHARD'S WELL   Mine
ST 4300 5552   L 11m VR 8m Alt 225m

Although technically on Axbridge Hill, this shaft lies close to, and at a similar level, to the Fry's Hill shafts. It was probably opened by ochre miners during the 19th century, who knew it as Capel's Hill Ochre Mine. Also known as Chardswell Cave, it was re-excavated by Clive and Janet North during the 1980s. The shaft is partly in Dolomitic Conglomerate and partly in a soft sandy deposit, and halfway down a squeeze enters a small chamber containing some exquisite coral deposits.

Ref: W. I. Stanton, Logbook 17 p 74 (1985)

CALLOW CAVE   Cave
ST 4350 5638   L 18m VR 3m Alt 200m

Also known as Callow Hill Cave, this cave is located almost 1 km due north of the Fry's Hill mast, below a cliff on the north escarpment of Callow Hill. A narrow rift and tight squeeze lead into a small chamber, presently occupied by badgers. It was excavated by Ray Mansfield in 1962, who recorded a few bones, fragments of medieval pottery, some old lead buttons and parts of a clay pipe. It was revisited by MEG in the 1980s, and along with neighbouring Callow Slit is a possible candidate for the 'rocky hollow' used by a Sidcot School Secret Society in the late 19th century as a kit stash during their unauthorised nocturnal trips to various west Mendip caves. **Callow Slit** (ST 4349 5637, L 3m VR 2m) is a small mined rift a few metres further west. It was dug briefly in 1971.

Ref: F.A. Knight, A History of Sidcot School – A Hundred Years of West Country Quaker Education 1808-1908, p 271, J. M. Dent & Company (1908) & R. Mansfield, Caving Diary and General Notes, Vol 1 p 38 (May 1962)

CHEDDAR WOOD

ELLENGE RISING
ST 4433 5456

Spring
Alt 28m

This small clear spring probably drains a prominent spur which comprises a large part of Cheddar Wood. It presumably once provided the water supply for a vanished dwelling marked on old maps as Ellenge Cottage.

Ref: W. I. Stanton, Logbook 16 p 28 (1983)

ST MICHAEL'S TUNNEL
ST 4408 5483

Tunnel
L 15m Alt 49m

This mysterious tunnel was exposed in 1983 during excavations to clear an old tennis court. A flat-out crawl yielded an artificial tunnel 2m high x 1m wide which terminated in a blank wall containing a curious borehole. It was probably part of an adit driven c.1878 in search of a clean water supply to feed the newly-built convalescent home. Evidently unsuccessful, it was quickly abandoned and the water was raised instead from a nearby well. Said to be 50m deep, this was infilled in 1979. St Michael's was founded by Matilda Blanche Gibbs of Tyntesfield, who had lost three children to tuberculosis. Inmates came from all over the country, the majority being under thirty and often destitute. Sadly, despite the excellent care provided, many of the patients died, and more than 800 lie buried in the overgrown churchyard located behind the house. Morality was clearly encouraged even after death as men and women are buried separately on either side of the large central cross. Nearby is a well-cared for Commonwealth War Graves Commission headstone, erected in memory of S. Gordon of HMS Pembroke (a shore barracks), who sadly perished here in 1919, presumably from wounds received during a German bombing raid on the facility in 1917 which killed 130 ratings. After Mrs Gibbs death, the Sisters of St Peter's ran the home until 1967 whereupon Group Captain Leonard Cheshire took over the running of the home to provide care for the incurably sick and disabled, a service which continues to this day. Leonard Cheshire was a truly remarkable man. The youngest group captain in the RAF, and one of the most highly decorated, he earned virtually every decoration available during WWII, including the Victoria Cross, the single highest award available for gallantry in the face of the enemy. He was a Wing Commander with the legendary 617 squadron (*The Dambusters*), carried out more than 100 bombing missions, and later flew as the British observer during the nuclear attack on Nagasaki. Cheshire founded his charity in 1948 and dedicated the rest of his life to helping disabled people. He also became famous for lecturing on conflict resolution and is sometimes attributed (incorrectly) with the legendary conversation prompted by a young lady who asked in all innocence if the Germans he encountered on one particular sortie '*were fokkers*'. Cheshire is alleged to have replied that '*the fokkers he was engaging were flying Messerschmitts!*' Sadly, the tale is apocryphal.

Ref: W. I. Stanton, Logbook 16 p 54-56 (1983)

FRY'S HILL PIT
Mine
ST 4355 5488
Alt 90m

Located just below the track, this large, fenced-off depression shows signs of mining activity and may be the remains of a collapsed chamber exploited by the ochre miners. Iron pyrites was said to occur in 'brassy lumps'. Burr records another site in this vicinity as Yew Tree Mine (ST 4346 5486), which may be the same one.

Ref: P. Burr, Mines and Minerals of the Mendip Hills, Volume I, MCRA p 98 (2015)

FIG TREE MINE
Mine
ST 4386 5511
Alt 93m

A low choked arch and some old walls are all that remains of this small ochre mine, which is located next to the track adjacent to an abandoned quarry. Burr records that large amounts of pyrites are present in the spoil heap, containing low concentrations of gold.

Ref: P. Burr, Mines and Minerals of the Mendip Hills, Volume I, MCRA p 98 (2015)

CHEDDAR WOOD OCHRE PITS
Mines
ST 4432 5492
L 7m VR 5m Alt 119m

The deepest of four small ochre pits situated in a line, this roomy fenced shaft yields only 3m of small unstable passage. The other holes consist of a tiny blind square pit located further down the hill, and two larger 3m deep pits higher up, the uppermost fenced. They were probably the last of the ochre pits worked above Axbridge, where operations finally ceased c.1937.

Ref: M. Clarke, N. Gregory and A. Gray, Earth Colours – Mendip and Bristol Ochre Mining, MCRA, p 142 (2012)

AXBRIDGE VILLAGE

HILLSIDE TUNNEL
Cave
ST 4306 5469
L 2m VR 1m Alt 35m

Located at ground level below an ivy-covered cliff alongside the busy A371, this low passage was presumably uncovered when the 'Strawberry Line' was driven through c.1869.

Ref: N. Barrington, W. Stanton, The Complete Caves and a view of the hills, p 29 (1977)

AXBRIDGE STATION WELL
Well
ST 4314 5469
L 16m VR 16m Alt 35m

This 2m diameter well, positioned beneath an old railway station platform, was exposed by workmen working on the Axbridge bypass in 1966. It was dived by the CDG to a depth of 3m, and yielded 3m of natural underwater passage, which proved too tight to follow. There was a faint current which may be connected to Axbridge Rising, located just 60m further south. The well is now buried beneath the bypass.

Ref: N. Barrington, W. Stanton, The Complete Caves and a view of the hills, p 29 (1977)

AXBRIDGE RISING
Rising
Alt 18m
ST 4312 5464

This large spring, referred to as the 'Great Spring' on a 13th century deed, wells up through Red Marl and gravel inside a masonry chamber and was the main water supply for the village for centuries. Also known as Fishponds, it feeds on into Axbridge Church Wells and once powered a water mill which stood here until the 18th century. The spring has a variable flow which is even known to fail in long dry spells. However in wet conditions it can grow quite powerful, as evidenced by the flood of December 23rd 2012, when collapse of a 600-year old enclosure released a local 'tsunami', which washed away gardens and mature trees.

Ref: N. Barrington, W. Stanton, The Complete Caves and a view of the hills, p 29 (1977) & Mendip Times Vol 10 Issue 3, p 43 (2014)

Axbridge Rising. Photo by Rob Taviner

AXBRIDGE CHURCH WELLS
Holy Well
Alt 16m
ST 4313 5459

These two wells, housed beneath late medieval chamfered arches, are supplied by the water from Axbridge Rising. They sit alongside the church and were restored during the latter part of the 19th century. There are four small satellite springs located beneath and to the west of Axbridge Square, which all deposited tufa during the Pleistocene.

Ref: N. Barrington, W. Stanton, The Complete Caves and a view of the hills, p 29 (1977)

CROWN INN OCHRE MINE
Mine
L 6m VR 6m Alt 16m
ST 4322 5456

Discovered in the back yard during the refurbishment of the Crown Inn in 2012, this small ochre working consisted of a 2m diameter shaft which terminated in a pool of water. It was quickly filled in.

Somerset Underground · Volume 2

SHIPHAM GORGE

Shipham Gorge is the name given to the deep valley which separates Shipham Hill from the east side of Callow Hill. Over the years the flanks have been subject to a significant amount of quarrying, and a number of small caves have been intercepted. The western flank is occupied by the large Callow Hill Quarry, which was established in 1919 by the Tiarks family (famous London bankers), ostensibly to provide employment for men returning from the trenches. This quickly gained a reputation for producing high quality lime, a part of the business that continued until 1969. Stone extraction, to the tune of a million tonnes a year, continues to this day, mainly to supply the raw material for an in situ concrete products plant. Callow means 'bare hill', and the flat ground on the summit was once the meeting place for Hannah More's famous Mendip Feasts, an annual event said to have attracted up to one thousand people from the local villages. Between 1823 and 1872, it even served as a race course. On the opposite side of the gorge several small quarries were developed during the 1880s. These quickly amalgamated into two larger concerns - the upper Shipham Hill Quarry, which mainly operated in the 1930s to supply material for the nearby A38, and the lower Shipham Gorge Quarry, which was one of most important concerns owned by Somerset County Council. The upper quarry ceased operations during the 1950s, and the council quarry in 1976, when it was used as a landfill site for waste accumulated from Callow Hill Quarry.

CALLOW HILL QUARRY

Several small caves have been recorded since quarrying began, including a deep stalagmite-lined rift which was revealed in 1936, then promptly covered over. Numerous others probably went the same way, although a few were examined before they were either buried or destroyed. The bulk of these are clustered around the high point of 'The Cutting', which links the two halves of the quarry. In 1979, Stanton examined a 3m deep mud-filled sinkhole above the north-west corner, and in 1988 examined a complex of fissures completely filled with Liassic sediments. Similar features are still visible in the western face. In 1939, a borehole sunk at ST 450 556 (BGS ST45NE9), intercepted a subterranean streamway in a cavity at roughly 150m depth, while a 100m deep well, sunk close to the office in 1964, proved completely dry.

Ref: W. I. Stanton, Logbooks 11 p 22 (1964), 15 p 2 (1979), 18 pp 60, 65, 66 & 68 (1988) & http://scans.bgs.ac.uk/sobi_scans/boreholes/386344/images/10707971.html

CALLOW ROCKS CAVE
ST 4477 5574

Cave
L 9m VR 4m Alt 220m

Situated in an obvious line of weakness, at the base of an overgrown cliff, this cave comprises a roomy phreatic tube with skylights. It still exists and was dug briefly by ACG in 1991 but has since been gated. Stanton christened it Callow Limeworks Cave, although old maps clearly show that Callow Rocks pre-existed the quarry.

Ref: W. I. Stanton, Logbook 4 p 130 (1949)

CAR PARK CAVE
ST 4462 5580

Cave
Alt 220m

A small decorated cave apparently uncovered in the car park by the quarry workshop during early 1980s. Explored by workmen, it consisted of a couple of chambers trending south towards the caves in 'The Cutting', which together may once have formed part of a single, larger system. No further details are known and the entrance was soon filled and lies buried somewhere beneath the quarry car park.

Ref: C. Dockrell, Personal communication

HONORARY HOLES
ST 4454 5581

Caves
L 7m VR 2m Alt 220m

Located opposite the workshop behind the honorary loading shed, this site comprised two superimposed cavities situated 3m apart. The upper hole (L 7m) was excavated by ACG in 1991, who described it as decorated with a possible way on, while the lower cave closed up after 4m. The upper hole still survives but is now covered with wire netting and partially choked with rubble.

Ref: C. Dockrell, ACG Logbook 7 p 82 (1991) & Personal communication

CUTTING CAVES & RIFT
ST 4456 5577

Caves
Alt 220m

Choked bedding planes were intercepted at floor level in 'The Cutting' leading to the main quarry in the early 1990s. Located on both sides, they were filled in almost immediately. In 1990 a rift appeared a little further south (ST 4458 5574). Estimated to be 10m deep, ACG descended it for 3m but abandoned it due to dangerous instability. It too was swiftly filled in.

Ref: C. Dockrell, ACG Logbook 7 p 7 (1990) & Personal communication

CALLOW QUARRY BEDDING PLANE DIG
ST 4453 5565

Cave
L 13m VR 4m Alt 220m

This bedding plane cave was uncovered by quarrying 1991. First mentioned as a sand-filled rift, it was examined by ACG, who pursued the bedding plane down to a choke and opened up a small blind rift in the roof. Although it still exists, the entrance was buried beneath rubble and now lies in an inaccessible position 20m above the quarry floor in the south face.

Ref: C. Dockrell, ACG Logbook 7 pp 57-58, p 92 (1991) & Personal communication

CALLOW LIMEWORKS QUARRY CAVE   Cave
ST 445 557   L 12m VR 6m Alt 210m
This hole appeared in the floor of the upper level in 1967 when a digger drove over it. A 6m deep shaft entered a low chamber floored with loose boulders. Beyond, a further 3m shaft led to an unstable choke. It was quickly filled in.

Ref: A Cave in Callow Rock Quarry, ACG N/L p 54 (Jun 1967)

CALLOW ROCK QUARRY CAVE   Cave
ST 4425 5564   L 2m VR 2m Alt 230m
In 1989, a blast shook open a roomy clay-filled hole, which appeared alongside the perimeter track, some 7m back from the south face. It has probably been quarried away, although stal and a large ochre deposit remain visible below.

Ref: C. Dockrell, ACG Logbook 6 p 123 (1989) & Personal communication

HAUL ROAD SHAFT   Cave
ST 4431 5596   L 4m VR 4m Alt 226m
An unstable natural shaft discovered by quarrying in 1990. It was visited by ACG and quickly filled in.

Ref: C. Dockrell, ACG Logbook 7 p 1 (1990) & Personal communication

NORTH FIELD CAVE   Cave
ST 4414 5630   L 19m VR 19m Alt 240m
This roomy shaft was intercepted during a northward expansion to the quarry in May 2019. It was destroyed shortly after. Large ochreous deposits have been recorded in this vicinity and subsequent investigations (based on geophysical tests) revealed a few clay-filled passages.

Ref: C. Dockrell, Personal communication

CROW CATCH   Sink / Cave
ST 4492 5590   L 12m VR 12m Alt 155m
Located on the edge of the quarry settling pond, this cave is accessed via an outsized blockhouse, constructed by the ACG during their dig in the late 1980s. A wet vertical rift terminates in unstable boulders. Work was finally abandoned due to concerns about flooding. The destination of the water is unknown (ranging from Axbridge to Cheddar), but given the underlying geology Ellenge Rising might be considered favourite.

Ref: C. Dockrell, ACG Logbook 6 pp 141-142 & 146 (1991)

SHIPHAM HILL QUARRY

ASH TREE PIT   Stone Pit
ST 4496 5614   L 2m VR 2m Alt 210m
This is a small circular rocky pit, surrounded by a debris cone. No obvious minerals have been found and as it clearly pre-dates the neighbouring quarry it was probably a stone pit dug to provide material for a nearby limekiln.

Ref: C. Dockrell, Personal communication

SHIPHAM HILL QUARRY RIFT — Cave
ST 4511 5617 L 4m VR 4m Alt 220m
Only accessible by using tackle from above, this open fissure is located near the top of the third tier of the quarry face.

SHIPHAM HILL QUARRY ALCOVE — Cave
ST 4509 5614 L 2m VR 2m Alt 220m
Located part way up the third tier of the quarry face, this small alcove can be reached by a steep loose scramble through undergrowth.

STOBART'S HOLE — Cave
ST 4521 5593 L 30m VR 9m Alt 175m
This loose roomy cavity, which extended across the south-eastern face of the lowest level of the quarry, appeared following a blast. Explored by BEC in 1971, there were drill holes in evidence and it was probably just a cavity separating the main face from a hanging wall of debris. It has since been completely removed.

Ref: W. I. Stanton, Logbook 12 p 70 (1971) & C. Dockrell, ACG Logbook 7 p 65 (1991)

SHIPHAM GORGE QUARRY

SHIPHAM GORGE QUARRY CAVE — Cave
ST 4520 5555 L 24m VR 17m Alt 150m
Several small caves have been recorded in the council-owned Shipham Gorge Quarry. In 1965, Stanton noted a phreatic tube situated high up the north-east face of the lower level. A workman told him that he'd crawled in 7m to the head of a drop. At least part of this cave still exists and is situated halfway up a 20m high cliff, marked by a small white tree. A phreatic tube has been followed upwards to a choke and it can only be reached using tackle. Ten years later, Stanton noted another entrance had appeared 5m below the floor level of the upper quarry, along with two others located just above floor level at the back of the upper level. These shallow phreatic cavities are still visible today in the south-east corner immediately below a large breached shaft filled with Triassic Keuper Marl (**Shipham Gorge Quarry Fissure**, ST 4526 5550 Alt 175m).

Ref: W. I. Stanton, Logbook 11 p 28 (1965), Logbook 13 p 117 (1975) & Logbook 14 p 25 (1976)

HOLLY WELL — Spring
ST 4516 5432 Alt 29m
Possibly corrupted from Holy Well, this spring is the source of the brook known as Holwell Rye. It is located at the head of a rubble-filled gully near the foot of Shipham Gorge.

Ref: W. I. Stanton, Logbook 16 pp 26-28 (1983)

BURRINGTON

Mankind's interest in the caverns of Burrington Combe date back over millenia, as evidenced by the Mesolithic remains uncovered in Aveline's Hole and finds covering a wide period in Rowberrow Cavern. The district's underground streams feed two great springs at Langford and Rickford, both of which undoubtedly served as important water sources for many centuries. The miners left their mark on Dolebury Hill and in Rowberrow Forest, as well as on both flanks of Burrington Combe, but it is as a mecca for caving that the combe is perhaps best known, especially for the local schoolchildren, thousands of whom first cut their teeth in the bowels of Goatchurch, Sidcot and Rod's. Many went on to have notable caving careers. Despite this intense weight of traffic, significant discoveries continue to be made, most recently by cave divers in Pierre's Pot in 2011 and then at Tween Twins Hole, a long-standing, if somewhat maligned dig that finally turned up trumps in 2016.

DOLEBURY AND ROWBERROW

Rowberrow was once home to Terry Pratchett, the author of the Discworld series of fantasy novels, who lived here from 1970 to 1993. Overlooked by an Iron Age hillfort, it is the site of a large Bronze Age barrow which was excavated in 1813. Dolebury Warren dominates the surrounding countryside and forms part of a Scheduled Ancient Monument and a 90.6 hectare biological SSSI. The defences and Celtic field systems date back to the 4th and 3rd century BC, and finds have included Bronze Age pottery, a spearhead, and a Roman coin. The medieval and post medieval rabbit warren provided a valuable source of meat and fur, and was completely enclosed inside the substantial ramparts. It is a particularly fine example of reuse of an existing structure, the banks and walls helping to prevent the rabbits from escaping and causing damage to surrounding farmland. Several long pillow mounds inside the warren are thought to have been constructed to encourage the rabbits to breed.

CHURCHILL CAVE   Cave
ST 4463 5909   L 6m VR 5m Alt 80m

Also recorded as Churchill Rock Shelter and located in a small overgrown quarry in the garden of The Manor House, this roomy phreatic cavity was discovered in 1865. It was originally filled with yellow ochreous earth. Old cave registry sheets note that it was '50 ft' long and the fill contained the bones of some small Pleistocene mammals. Local legend holds that WWII explosives were sealed inside the rear of the cave. Ochre was also worked below the northern rampart of Dolebury hillfort c.1880 and one large conical depression contains ochre and stalagmite.

Ref: N. Barrington, W. Stanton, The Complete Caves and a view of the hills, p 132 (1977), M. Clarke, N. Gregory and A. Gray, Earth Colours – Mendip and Bristol Ochre Mining, MCRA, p 130 (2012) & N. Richards, Personal communication

ROWBERROW WATERWORKS TUNNEL   Tunnel
ST 4452 5885   L 450m Alt 100m

Built in 1922 by the Bristol Waterworks Company, the tunnel carries a large cast iron pipe and was driven to supply fresh water to Sandford and to Blagdon and Chew Valley lakes. The tunnel is securely locked but rails can be seen curving off into the distance. There are several interesting solutional holes in the adjacent quarry, although none are penetrable. The tunnel was heavily guarded during WWII, sparking rumours that the Crown Jewels were stored within, a claim which the company has always denied. However such rumours are not surprising given the amount of valuable material that was stored in the region during this period, including several priceless works of art in Westwood Quarry near Bradford-on-

Avon, and the Domesday Book in Shepton Mallet prison. As to their true fate, in 2018 it was finally revealed that the jewels had been hidden inside a biscuit tin, in chambers buried beneath the entrance to Windsor Castle!

DOLEBURY CAVERN
Lost Cave

There are three distinct references to an apparently lost cave at Dolebury. The first, purportedly located somewhere below the hillfort fortifications, was mentioned in an unpublished manuscript written by the antiquary, John Strachey in 1736. He described it as '... *under this fortification is an hole or Cave called Guy Hole, altogether as remarkable as that at Woky but the former being near a City & this remote from any place of Entertainment is not often visited by Travellers ..*'. Investigations by Irwin concluded that it was an alternative name for Goatchurch Cavern. On the 21st May 1819, a day of constant rain, the antiquarian Rev. John Skinner walked with a Mr. Jelly to visit the warrener at Dolebury Hillfort. During a conversation, the warrener gave Skinner '*an account of a prodigious excavation about a mile distant which was discovered while sinking for lead; it was of such extent that, he says, when he threw in a stone as far as he could he could not here it touch the sides.*' Sadly no further clues are forthcoming and it could lie in either direction along the northern outcrop of the limestone between Burrington and Sandford. However the dimensions of the shaft are so vast that it may be an oblique reference to the legendary Sandford Gulf. Skinner later recorded details of a lead mine adit at Dolebury, which is almost certainly Dolebury Levvy. This adit was mentioned by Knight, in Heart of Mendip, who outlined details of a horizontal gallery driven during the period 1829-1831. Subsequently blocked, it seems that this was sometimes referred to as Dolebury Cavern. Knight also records a deep mine shaft located midway from the fort to the eastern end of the hill (probably Dolebury Warren Mineshaft) which he speculated was opened up for lead.

Ref: D. J. Irwin, The Lost Caves of Mendip, BEC Belfry Bull 505 pp 31-46 (1999)

DOLEBURY LEVVY
Mine

ST 4516 5873 L 168m VR 8m Alt 110m

Also known as Dolebury Cavern, Dolebury Adit and Dolebury Lead Adit, this horizontal working was originally excavated c.1829-1831 in search of haematite. A short, descending flat-out slide gains a mined circular tunnel, 2m high x 2m wide, which can be followed until it ends abruptly at a solid rock wall. In 1904, there was an unsuccessful attempt to mine ochre commercially and its final use was as an Auxiliary Unit Operational Base during WWII. The mine was reopened by ACG and the Society for Mines Research in the Bristol Area in 1969. Trial pits were also sunk on the southern ramparts of the hillfort, close to Yew Tree Cliff.

Ref: M. Clarke, N. Gregory and A. Gray, Earth Colours – Mendip and Bristol Ochre Mining, MCRA, p 129 (2012) & Mendip Underground, MCRA p 111 (2013)

DOLEBURY HILL CAVE
ST 4519 5874   **Cave**
L 9m VR 2m Alt 115m

Located at the foot of low bluff, 15m up the steep bank above the path, this small cave consists of a low creep which soon becomes too tight to follow. There is pile of digging spoil next to the entrance and the cave appears to be frequented by animals.

Ref: C & C Dockrell, Additions to the Complete Caves of Mendip, ACG N/L pp 1-2 (Jun 1992)

THE PARALLEL GASHES
Gashes!

Visible on old aerial photographs, a series of parallel gashes can be traced from inside the hillfort as far east as the small plantation containing Dolebury Warren Mineshaft. These are now filled in, but of particular interest is a note left by Great Elm resident Thomas Morgans, the eldest son of the famous mining engineer Morgan Morgans. *'The writer has been informed that in the limestone of Dolebury Warren…two parallel gashes, several feet apart and both filled with clay, had been bottomed at a depth of 30 to 40 feet, and that in both cases the bottom was concave transversely in solid limestone. This suggests that these two gashes were ancient watercourses and that they had been entirely excavated by water.'* The 'gashes' were probably neptunian dykes.

Ref: T. Morgans, Notes on the Lead-Industry of the Mendip Hills. Transactions of the Institute of Mining Engineers, Vol 20 pp.478-494 (1900-1901)

DOLEBURY WARREN MINESHAFT
ST 4581 5881   **Mine**
L 5m VR 5m Alt 166m

Situated in a clearing in a small fir tree plantation, roughly halfway along Dolebury Warren, nothing can now be seen of a 5m deep mineshaft, recorded by ACG in 1992. There are several choked mineshafts in this vicinity, which are all that remains of the mines that once peppered the summit of this hill. Knight, in his Heart of Mendip, mentions a deep lead mine, apparently situated midway between the fort and the eastern end of the hill, which Skinner had earlier reported to be 30 fathoms deep. Skinner probably visited the site based on the information he received from the warrener at Dolebury hillfort watchtower, who described a *'prodigious excavation about a mile distant which was discovered while sinking for lead; it was of such extent that, he says, when he threw in a stone as far as he could he could not here it touch the sides.'* One of the shafts was reputed to have disturbed a hoard of gold and silver Saxon and Roman coins, whereabouts unknown.

Ref: C & C Dockrell, Additions to the Complete Caves of Mendip, ACG N/L pp 1-2 (Jun 1992)

DOLEBURY WARREN SWALLET
ST 4553 5861   **Sink**
L 2m VR 2m Alt 110m

This sink, located close to the path that runs along Dolebury Bottom, was dug in alluvium by ACG in 1988. No solid rock was seen and prospects were described as poor. In wet weather the swallet frequently overflows and is augmented by water rising from a small spring 30m further south-west (ST 4550 5859).

ROWBERROW BOTTOM SHELTER
Cave
ST 4537 5847
L 3m VR 2m Alt 114m

A regular hobbit hole, this attractive little shelter is situated among the ruins of a tiny abandoned hamlet a few metres above the path that follows the valley floor. Surrounding walling suggests that it may once have been incorporated inside a dwelling, probably for use as storage, for keeping animals, or even for habitation.

Ref: N. Barrington, W. Stanton, The Complete Caves and a view of the hills, p 132 (1977)

ROWBERROW SWALLET
Sink
ST 4553 5809
Alt 128m

This invisible sink swallows part of the perennial stream that gradually vanishes underground in Rowberrow Bottom. In 1987, tests proved that the water passed through Aldermaston Chamber in Mangle Hole, en route to Banwell Spring. This destination had been tested by MKHP, in conjunction with the University of East Anglia, in 1975 (7 days). At one time the swallet was concreted over, before being reopened by indignant locals, who understood the important contribution it made to flood prevention. Water sinking in this valley is frequently compromised by muddy runoff and is thought to be one of the major causes of turbidity in the Banwell Spring.

Ref: N. Barrington, W. Stanton, The Complete Caves and a view of the hills, p 133 (1977)

SWAN INN SWALLET
Sink
ST 4556 5786
L 3m VR 2m Alt 140m

Under flood conditions water from the surface stream sinks into a 3m deep pit covered by an inspection hatch. The site was dug briefly by SSSS in 1961, and in 1977 Bristol Waterworks tested the water to Banwell Spring (37 hours). This confirmed another test undertaken by WCC who recorded a time of 50-150 hours. In 1979, council workmen excavated two further sinks, located 20m north and 30m south of the swallet which they subsequently filled with rubble. Further clearing work c.1994, accidentally opened up yet another swallet which took all the flow until it became blocked during winter floods.

Ref: W. I. Stanton, Logbook 14 p 153 (1979)

ROWBERROW CAVERN
Cave
ST 4598 5803
L 25m VR 6m Alt 177m

Rowberrow Cavern is a Scheduled Ancient Monument, highly regarded for the rare Palaeolithic hearth material discovered both inside and outside the cave. An abandoned resurgence, it is formed in Dolomitic Conglomerate and has an extensive platform outside, which probably represents a collapsed extension of the cave. Beyond the impressive entrance, the cave rapidly

Rowberrow Cavern. Photo by Rob Taviner

narrows, reaching a solid choke after 25m, with some spoiled formations and one short side passage on the left. Deep excavation trenches are the results of archaeological work carried out by UBSS in 1920-26, who unearthed activity covering a wide period, including hearth material dating from the Upper Palaeolithic. There were indications of Mesolithic activity, plus a range of Neolithic/Early Bronze Age material, including flint implements, knives and Beaker potsherds. Iron Age finds included bone fragments, pottery, a light blue bead, and some traces of iron smelting, while Romano-British finds included potsherds, coins, and a few pieces of copper alloy and lead. Important material deeper into the cave remains untouched. In 2016, small stalagmites were found nearby lying among the rubble of some grubbed out forestry workings (ST 4590 5803). This strongly suggests that there may once have been other caves in the immediate vicinity.

Ref: English Heritage List Entry Number: 1011926

ROWBERROW FOREST MINE Mine
ST 4592 5750   L 4m VR 2m Alt 217m
Located in woodland on the western slopes of the forest above Holloway Rocks, this tiny opening in a line of small pits yields a narrow descending calamine working. It is largely choked with brushwood.

Ref: N. Richards, Personal communication

ROWBERROW FOREST SWALLET Sink
ST 4663 5744   Alt 260m
Staking a claim for Mendip's shortest surface stream, a tiny spring vanishes almost immediately into a wooded swallet. Sinking water has been noted elsewhere in this part of the forest (notably in the trackside drainage ditches),

and the immediate area was also worked for iron, as evidenced by a number of pits (**Rowberrow Forest Iron Pits**) and a 2m deep x 3m long open slot, situated 100m north-east of the swallet.

Ref: N. Richards, Personal communication

WATER VALLEY SWALLET
ST 4670 5677

Sink
L 6m VR 3m Alt 245m

Located near the south-east corner of the forest, 400m north-west of Tyning's Farm, this overgrown swallet was inconclusively dug c.1940. The water probably joins the stream flowing through the adjacent miniature ravine, which rises nearby and continues down into Rowberrow Bottom.

LANGFORD

SAYE'S LANE RISING
ST 4570 5954

Spring
Alt 43m

A small marshy pond that can overflow during wet weather and waterlog the immediate surroundings.

Ref: N. Barrington, W. Stanton, The Complete Caves and a view of the hills, p 141 (1977)

LANGFORD RISING
ST 4661 5929

Rising
Alt 40m

Tests conducted by UBSS suggests that the flow from all of the Burrington stream sinks unite into a single conduit before diverging to flow to two major risings situated adjacent to the A368. Situated at the foot of Mendip Lodge Wood, this isolated pool is the smaller of the two and is the source of Langford Brook. Also known as Langford Spring or Hylisbrook (Hilsbrook) Spring, it rises through gravel overlying Red Marl, and was mentioned in an Anglo-Saxon charter of AD 904 when it was recorded as the 'Great Spring'. Once an important source of water to the Bristol Waterworks Company, who built the attractive 'Gauge House' located immediately downstream', 'proved' feeders include Bath Swallet, Read's Cavern, Flange Swallet, Pierre's Pot (sourced from East Twin Swallet, Goatchurch and Flange), Yew Tree Swallet (via Goatchurch Cavern), East Twin Swallet and Ellick Farm Swallet. The results are based on several different tracing attempts (not all successful), and the wide variations in the recorded times render them almost meaningless. The rising responds rapidly and frequently to flood events, but equally can dry up altogether, which suggests that a good fraction of its water is derived directly from surface swallets. The fact that swallet water common to both risings routinely arrives at Langford Rising later, has led some commentators to suggest that it may be in the process of capturing Rickford Rising.

Ref: N. Barrington, W. Stanton, The Complete Caves and a view of the hills, p 106 (1977) & A. R. Farrant, G. J. Mullan & A. A. D. Moody, Speleogenesis and Landscape Development in the Burrington Area, Somerset, UBSS Proc 24 (3) pp 207-252 (2008)

Langford Gauge House. Photo by Rob Taviner

BURRINGTON COMBE

Burrington Combe is one of Mendip's most important limestone gorges and an SSSI. Numerous caves have been explored, several of which have been occupied by humans, including Aveline's Hole, which contained the earliest scientifically dated cemetery in Great Britain (9114 BP to 8860 BP, plus or minus 100 years), along with a series of rare inscribed crosses believed to date from the early Mesolithic. The most famous cave however is Goatchurch Cavern, a perfect novice cave which was first recorded in 1736. Extensively explored by lead miners in the 19th century, later excavations uncovered the bones of mammoth, bear, hyaena and cave lion dating from the Pleistocene period, plus several ritual protection marks, or 'witch marks' as they are more commonly known. It was later an unsuccessful show cave. Above ground, the combe is most famous for its association with the Rev. Augustus Toplady, who allegedly composed the famous hymn 'Rock of Ages' while sheltering inside a cleft from a thunderstorm. The hymn is regarded as one of the four great Anglican hymns, and was a favourite of Prince Albert, who even asked for it to be played on his deathbed. Sadly, the story now appears unlikely, for Toplady appears to have composed the hymn sometime after he left the area and the attribution of the hymn to the rock that bears its name only seems to have arisen well after his death. The combe was also the scene of an early cave fatality (1875), when the unfortunate Joe Plumley expired while being pulled up from the shaft which now bears his name. In 1944, as D-Day approached, Mendip was declared a military exclusion zone, home to both a vast ammunition dump and the US 9th Army. Prior to the embarkation to Normandy the combe was lined with a column of camouflaged military vehicles which stretched from the bottom of the combe all the way back to the Miners Arms!

LOST CAVE OF BURRINGTON I   Lost Cave

A mysterious lost cave, which featured in the Gentleman's Magazine of 1805 and was said to contain the remains of a hundred human burials. Research by Desmond Donovan and George Boon in 1954 showed it to be 'Fairy Toot', a Neolithic chambered tomb near Nempnett Thrubwell (ST 5205 6178), which had been investigated in 1789 by the Rev. Thomas Bere. In a catastrophic act of vandalism the tomb was later largely destroyed. Writing in 1929, Wade and Wade described *'a remarkably fine tumulus of masonry, said to have been one of the finest in Britain, in the chambers of which skeletons have been discovered. A few vestiges of it now only remain, the rest has been used as a lime-kiln.'* Unusually the site was also known as a place for curing warts!

Ref: D. J. Irwin, The Lost Caves of Mendip, BEC Belfry Bull 505 pp 31-46 (1999)

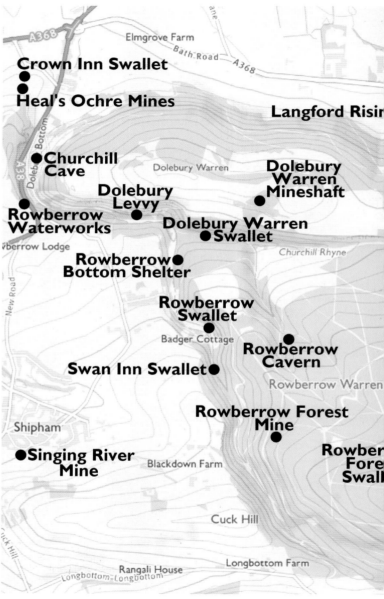

Burrington · Burrington Combe

Rowberrow and Burrington

Contains Ordnance Survey data
© Crown copyright and database right 2018

LINK LANE & MENDIP LODGE WOOD

Mendip Lodge was built by T. S. Whalley, a man who made his somewhat dubious fortune through marrying rich heiresses. The house had an Italian verandah over 30m long and there were up to 52 grottoes in the garden, the remains of some of which can still be found buried beneath the thick rhododendrons. Said to have been the envy of the district, the house fell into disrepair when Whalley's luck, and the money, finally ran out. Other than the cellars, little now remains of the house itself, and the best surviving feature is probably 'The Lookout', a recently restored walled enclosure, which commands fine views across the Congresbury Yeo to Broadfield Down and the Bristol Channel.

LOST CAVE OF BURRINGTON 4 — Lost Cave

A lost cave reported by E. K. Tratman and said to be located in a shallow valley to the south of The Link. It was filled in almost immediately. The precise location is uncertain (especially as there is no shallow valley to the south of The Link), but it may well be the deep 'old artificial shaft' noted by Stanton and still visible as a collapse in the floor of Mendip Lodge valley at ST 4742 5874. In a small quarry further down the valley (ST 4746 5896), there is a low undercut, greatly concealed by brush.

Ref: D. J. Irwin, The Lost Caves of Mendip, BEC Belfry Bull 505 pp 31-46 (1999) & W. I. Stanton, Logbook 2 p 76 & 90 (1945)

ASH TREE RIFTS — Mines

ST 4751 5866
L 3m VR 2m Alt 151m

Located right at the top of the steep west bank of Burrington Combe, almost immediately opposite Aveline's Hole, these two short parallel open rifts, formed in Dolomitic Conglomerate, are best approached from Link Lane. Signs of mining at the same level a short distance to the north, suggests that they are trial iron pits. This site was correctly described but incorrectly identified as Nameless Cave (see below) by Stanton in 1970.

Ref: N. Barrington, W. Stanton, The Complete Caves and a view of the hills, p 116 (1977)

P HOLE — Cave

ST 4726 5846
L 5m VR 3m Alt 172m

This short descending passage, located on the eastern side of the Bath Swallet depression, was probed by the UBSS in 1946. There is also a 5m deep pit a few metres to the south that subsides at regular intervals. This is a lead mine, and there are other mines nearby. The accumulation of mineral waste suggests that the depression was used for ore washing, and the stream actually derives from a long leat, dug by miners from the Hunter's Brook.

Ref: N. Barrington, W. Stanton, The Complete Caves and a view of the hills, p 36 (1977) & N. Richards, Personal communication

BATH SWALLET
Cave
ST 4725 5845 L 220m VR 41m Alt 172m

This cave, originally referred to as Bathing Pool Swallet or Bath Hole, was named after the brick-lined bath constructed in the streambed in 1921 by UBSS members whose caving hut lay nearby. Two short climbs and a 15m deep pitch yield a large phreatic passage leading to a complex network of passages. One connects with Rod's Pot via a 6m upwards pitch. Digging by UBSS between 1944 and 1946 opened up a short section of passage, but the majority of the cave was discovered by WCC in 2001. The link with Rod's Pot was opened by ChCC in 2007. UBSS dye-tested the large but intermittent stream to Langford Rising in 1962.

Ref: Mendip Underground, MCRA pp 61-64 (2013)

ROD'S POT
Cave
ST 4721 5843 L 300m+ VR 46m Alt 174m

Originally called Pearce's Pot (in honour of UBSS club secretary Dr Rodney Pearce), this popular cave features a steeply-descending canyon leading to the head of two blind pots. Beyond, the roomy Main Chamber leads to a 6m pitch and the connection with neighbouring Bath Swallet. There are numerous side passages, pitches and entertaining loops connecting Main Chamber with the two blind pots. The cave was opened by UBSS in 1944, after only four hours digging. ChCC greatly extended the cave and linked it to Bath Swallet in 2007. Work continues.

Ref: Mendip Underground, MCRA pp 278-281 (2013)

DRUNKARD'S HOLE
Cave
ST 4714 5839 L 147m VR 48m Alt 175m

Also recorded as Cow's Hole, a narrow, steeply-inclined passage yields a series of parallel rift chambers, with a 10m deep pitch. The cave was discovered about 1923 by UBSS, following a night out at The Swan at Rowberrow, and largely extended to its present length by WCC in 1989.

Ref: Mendip Underground, MCRA p 112 (2013)

BOS SWALLET
Cave
ST 4709 5836 L 78m VR 42m Alt 178m

A steeply-descending swallet cave with two short pitches (7m and 5m). SSSS opened the site between 1946 and 1948, finding flints, pottery and bones, and UBSS have re-examined the site on a number of occasions, uncovering a leaf-shaped arrowhead from the Early Neolithic and a hearth, flints and pottery sherds dating from the Beaker period. A pit and burnt mounds from a late phase of activity have been interpreted as a Middle Bronze Age boiling site. A new UBSS dig, commencing in 1994, extended the cave to its present limit.

Ref: Mendip Underground, MCRA p 71 (2013)

FOX HOLES
Cave
ST 4681 5844 L 6m VR 1m Alt 163m
Also known as Foxholes Swallet, Spider Holes and Swallet D, these are a series of small tunnels, digs and open bedding planes situated a few metres west of Read's Cavern. The longest and most prominent is situated directly above the main cave's stream entrance, while similar small holes to the east of Read's Cavern appear to be inhabited.

Ref: N. Barrington, W. Stanton, The Complete Caves and a view of the hills, p 81 (1977)

READ'S CAVERN
Cave
ST 4682 5844 L 1225m VR 63m Alt 161m
Originally called Keltic or Celtic Cavern, this cave was re-christened by the Ordnance Survey in honour of Reginald Read, one of the original explorers. The cave has two entrances, which lead directly into the large Main Chamber. Below, an extensive network of narrow passages can be followed through a vast, unstable boulder ruckle to reach a final impenetrable sump. The cave was opened in 1919 by UBSS and has been extended at various intervals since. Painted gridlines and numbers on the rear wall, and several boulders in the Main Chamber, were drawn in connection with archaeological excavation by UBSS in the 1920s. Much pre-Roman Iron Age material was recovered, including some human remains. Further work by UBSS in 2010, suggests that the cave may have been used more for ritual purposes than for habitation or industrial use. A brooch, found within a charcoal rich Iron Age deposit, is continental in style, and has been dated to the La Tene period (450-20 BC). MKHP spore tested the stream (Hunter's Brook) to both Langford Rising and Rickford Rising in 1968.

Ref: Mendip Underground, MCRA pp 252-259 (2013)

Read's Cavern. Luke Devenish collection. Photo courtesy of WCC

BURRINGTON COMBE (MAIN VALLEY)

MILLIAR'S QUARRY CAVE
Cave
ST 4772 5898 L 17m VR 14m Alt 100m

Milliar's Quarry lies on the east bank of the combe and is approached via a track which runs behind the Burrington Inn. The cave is located 4m above the floor on the right-hand side of the quarry and consists of a small, steeply-descending tube which follows the bedding. It was first entered by UBSS in 1951.

Ref: N. Barrington, W. Stanton, The Complete Caves and a view of the hills, p 115 (1977)

CAFE HOLE
Cave
ST 4762 5889 L 1m VR 1m Alt 92m

This tiny rift, in Dolomitic Conglomerate, lies at road level on the west bank directly opposite the Burrington Inn car park.

Ref: N. Barrington, W. Stanton, The Complete Caves and a view of the hills, p 48 (1977)

PLUMLEY'S HOLE
Cave
ST 4766 5875 L 20m VR 20m Alt 100m

A shaft discovered by quarrying in mid-December 1874 down which stones were heard to drop for 3-4 seconds. On the 5th January 1875, Joseph Plumley, a 54-year old local labourer, descended a narrow shaft said to be 150ft deep, where he reached a stream. However other accounts quote 70ft, 120ft and more enigmatically *'beyond all sound of his voice.'* One particularly flowery account recorded 'the cave is of immense depth and full of chambers branching off in all directions ... to give a slight idea of depth, a stone thrown down can be heard for 10 seconds!' Sadly, Plumley died during the ascent to the surface, and his body wasn't recovered until the following day when brave George Clarke discovered him doubled-up beneath a projecting ledge at 70ft depth with the rope apparently tangled up around his neck. The cause of death was recorded as accidental hanging, doubtless the result of the vigorous pulling of several men trying to haul him to the surface. The shaft was later filled with many cartloads of rubble and the entrance plugged with a tree stump. Balch claims to have descended the shaft for 60ft during the 1890s, and when UBSS excavated here in 1919-1920, Tratman recorded a depth of 40ft. A small ledge 30ft down carried a small stream and this led to speculation that this was the point where Plumley must have perished for his feet were said to have been found dangling in a stream. The shaft was sealed with concrete in 1924, briefly reopened, then sealed permanently in 1946. It is located in the obvious rock alcove opposite the Rock of Ages.

Ref: A. F. Dougherty, D. J. Irwin, C. Richards, The Discovery of Plumley's Hole, Burrington Combe and the Death of Joe Plumley, UBSS Proc 20 (1) pp 42-58 (1994)

Burrington Combe and surrounding area. Photo by Keith Savory

Burrington · Burrington Combe

NAMELESS CAVE
Cave
ST 4756 5868
L 8m VR 3m Alt 130m

This cave is located on the steep west bank of Burrington Combe, immediately above the cliff face opposite Aveline's Hole. A roomy archway leads directly to a second chimney entrance with a low extension to the east. It was first recorded by Tratman in 1963 but subsequently misidentified by Stanton, who confused it with Ash Tree Rifts (see above).

Ref: E. K. Tratman, The Hydrology of the Burrington Area, Somerset, UBSS Proc 10(1) pp 22-57 (1963)

AVELINE'S HOLE
Cave
ST 4761 5867
L 70m VR 16m Alt 99m

Also recorded simply as The Cave and obvious to all visitors to Burrington Combe, the impressive entrance to Aveline's Hole leads directly into a large passage which descends steeply to a locked steel gate. This was installed in 2003 to protect a rare engraving, probably of Mesolithic date, which was found etched on the wall some 30m into the cave. Beyond the gate the main passage descends to a muddy choke, while a crawl on the left just before the gate yields a tiny series of passages and avens which terminate close to the surface. The cave, which is probably an ancient resurgence, was discovered in January 1797, when two men pursued a rabbit into a small hole (at the top of the present entrance arch). They discovered a number of human skeletons, some encased in tufa, on the cave floor. The mud-filled shaft at the end of the cave was dug by William Boyd Dawkins in the 1860s, and later archaeological work, mainly by UBSS in the 1920s, has shown this site to be the earliest Mesolithic cemetery in the country. The bones of at least twenty individuals have been found, dating from between 9114 to 8860 BP, plus or minus 100 years. Two of the burials had a nest of seven ammonites interred alongside them, ornamental fossils traditionally imbued with magical properties. Among the later finds were a six-barbed antler harpoon and a necklace of perforated shells. The cave, which is a Scheduled Ancient Monument, is of wider European significance and is named after William Talbot Aveline, a noted geologist and mentor of William Boyd Dawkins, who visited the cave and was awarded the prestigious Murchison Medal in 1894. During WWII the cave served as a makeshift shelter for families fleeing the German blitz on the surrounding towns and cities.

Ref: Mendip Underground, MCRA p 42 (2013)

LOST CAVE OF BURRINGTON 2
Lost Cave

A lost cave apparently located 50 yards distant from Aveline's Hole in which a bronze axe head was found on a side ledge 28 feet below the surface. Reported in the Gentleman's Magazine of 1805 the cave was probably quarried away shortly after, although Supra-Aveline's Shaft remains a possible contender.

Ref: D. J. Irwin, The Lost Caves of Mendip, BEC Belfry Bull 505 pp 31-46 (1999)

Aveline's Hole. Photo by Mark Burkey

SUPRA-AVELINE'S SHAFT
Cave
ST 4768 5860 L 7m VR 5m Alt 119m

Also known as Nameless Hole, this tiny cave, which is practically invisible from below, is situated in the middle of the quarry face, 85m up-combe from Aveline's Hole and 15m above the level of the lay-by. Two small entrances yield a narrow and near vertical tube. Given its location, it must be a possible candidate for Lost Cave of Burrington 2.

Ref: N. Barrington, W. Stanton, The Complete Caves and a view of the hills, p 152 (1977)

QUARRY CAVE
Cave
ST 4769 5852 L 2m VR 1m Alt 115m

This east bank cavity lies within a steeply-sloping bedding plane exposed in the north-east corner of a small disused quarry located among scrubland a short distance up-combe of the larger quarry immediately above Aveline's Hole. There is a tiny waterworn fissure at the rear.

200 YARD DIG
Cave
ST 4767 5850 L 6m VR 6m Alt 113m

Also known as Anon 2, only the top metre or so now remains of the narrow steeply-inclined shaft sunk by UBSS in 1920-22, which yielded only an impenetrable waterworn rift. It lies beneath a small rock face on the east bank close to the road, a short distance up-combe from Quarry Cave and approximately two hundred yards from Aveline's Hole (hence the name). It was reopened by persons unknown c.1970 and refilled shortly after.

Ref: N. Barrington, W. Stanton, The Complete Caves and a view of the hills, p 165 (1977)

TUNNEL CAVE
Cave
ST 4763 5829 L 10m VR 4m Alt 157m

This cave, which is probably best approached from above, lies on an overgrown ledge at the northern (downhill) end of a steeply-inclined cliff high up on the west bank of the main combe above the junction with the West Twin Brook. The cave has two entrances which lead to a short section of small phreatic passage.

Ref: N. Barrington, W. Stanton, The Complete Caves and a view of the hills, p 164 (1977)

TUNNEL BEDDING PLANE
Cave
ST 4763 5828 L 1m Alt 163m

Also known as Whitcombe's Hole 2, this choked bedding plane is located at the foot of the steeply-sloping cliff, approximately midway between Whitcombe's Hole and Tunnel Cave.

Ref: C. Howell, D. J. Irwin, D. Stuckey, A Burrington Cave Atlas, BEC Cave Report (1973)

WHITCOMBE'S HOLE
Cave
ST 4763 5827 L 10m VR 2m Alt 166m

Located near the crest of the west bank of Burrington Combe, above Tunnel Cave, and at the opposite (upper) end of the steeply-inclined cliff, this important site is possibly best approached from West Twin Brook. It consists of a small yet roomy phreatic tunnel, which was first excavated by Boyd-Dawkins c.1860, who retrieved animal remains plus potsherds and an implement described as 'angle iron' - possibly a Roman coffin binding. The pottery was initially regarded as Roman, but recent research suggests an Iron Age date for the deposit. Sadly the 'angle iron' is now lost but the other finds reside in Taunton Museum. An archaeological investigation by Vince Simmonds in 2011 recovered 19th and 20th century glass, plus a number of small mammal and bird bones and a single worked flint flake. The precise identity of the Whitcombe who gave his name to the cave remains unknown.

Ref: N. Barrington, W. Stanton, The Complete Caves and a view of the hills, p 173 (1977) & V. Simmonds, An overview of the archaeology of Mendip caves and karst, www.mcra.org.uk, (2014)

TWEEN TWINS HOLE
Cave
ST 4774 5827 L 251m VR 18m Alt 150m

Often referred to as Fester Hole, this fascinating cave is located at the foot of a steeply-sloping cliff on the south bank and comprises no less than four entrances (three currently accessible). Initially excavated by CSS in 1967, and dug intermittently by WCC since 1985, the cave consists of a complex of roomy and well-decorated phreatic tunnels, the majority of which were entered in 2016. To protect the vulnerable formations the cave is gated and a leadership system is in operation.

Ref: N. Barrington, W. Stanton, The Complete Caves and a view of the hills, p 164 (1977)

BARREN HOLE
Cave
ST 4781 5823 L 3m VR 2m Alt 165m

This tiny cave lies at the foot of a steeply-inclined rock outcrop located on the south bank of the combe high above the road. First noted by UBSS 1919-1920, it comprises a tiny chamber and a small ascending aven. The curious stone pillar in the entrance, which Stanton rather tongue-in-cheek considered might be an image of a local sage, remains in situ, but appears to be entirely natural.

Ref: N. Barrington, W. Stanton, The Complete Caves and a view of the hills, p 36 (1977)

BRUCE'S HOLE
Cave
ST 4786 5830 L 1m VR 1m Alt 143m

Located at the foot of a low cliff high up on the north bank of the combe, this site was briefly investigated in 1970 but revealed no way on. It was apparently named after SWCC member Bruce Foster. Only a low arch filled with scree remains.

Ref: N. Barrington, W. Stanton, The Complete Caves and a view of the hills, p 47 (1977)

Tween Twins Hole. Photo by Pete Hann

GOON'S HOLE
Cave
ST 4791 5825 L 25m VR 10m Alt 136m

This site was first dug by the BEC, who christened it Snogging Hole, after Keith 'Snogger' Hawkins. It has also been known as Hawkins' Hole, Snogging Pot and Burrington Hole 2. Subsequently lost, it was 'rediscovered' by Alan 'Goon' Jeffries of the SMCC & GSG, who stumbled upon it accidentally when he stopped off to take a leak in the bushes! The tight, scree-filled entrance, which lies 50m down-combe from Lionel's Hole on the north bank right next to the road, consists of a series of tight wet crawls terminating in a small bedding chamber, which must lie very close to Lionel's Hole, where the water reappears. The dig was pushed to a conclusion by WCC between 1986 and 2008.

Ref: N. Barrington, W. Stanton, The Complete Caves and a view of the hills, p 87 (1977)

LIONEL'S HOLE
Cave
ST 4796 5823 L 1000m VR 43m Alt 141m

Originally known as Burrington Hole, Lionel's Hole is located on the north bank, a short distance down-combe of the junction with the East Twin Brook. It offers a sporting round trip, notable mainly for the maze-like complexity of 'The Labyrinth', which provides a fine test of route-finding ability. The small stream seen in the lower reaches possibly derives from sinks in the Ellick Farm area to the east and may follow a parallel, more northerly route to the one seen in Pierre's Pot en route to Langford and/or Rickford Rising. The entrance boulder ruckle was dug by several teams before WCC finally succeeded in entering the cave in 1970. The further reaches, including the round trip, were excavated by members of the same club in the 1970s and 80s. Recently there have been claims made that the cave was actually first entered by MNRC members Lionel Haines (after whom the cave is named) and John Letheren in 1967. However the WCC team noted no earlier visitors and contemporary documentation (including log entries written by John Letheren himself in 1970) clearly contradict the claim.

Ref: Mendip Underground, MCRA pp 192-195 (2013)

LAY-BY DIG
Cave
ST 4802 5818 L 3m VR 3m Alt 144m

Also known as Anon 1, nothing now remains of this small shaft located at road level immediately behind the lay-by on the south bank a short distance up-combe from the junction with the East Twin Brook. Dug open by the BEC 1970-71, it was filled in following instructions from the landowner. At the western end of the lay-by there is a walled enclosure which swallows drainage from both the road and the opposite flank of the combe. Prior to the clearance of East Twin Swallet the valley stream often sank at this point.

Ref: N. Barrington, W. Stanton, The Complete Caves and a view of the hills, p 106 (1977)

Traverse in Lionel's Hole. Photo by Matt Voysey

1921 DIG   Cave
ST 4814 5821   L 1m VR 1m Alt 171m
Concealed among bushes 60m west of Trat's Crack, this small alcove was excavated by UBSS in 1921-22.

Ref: N. Barrington, W. Stanton, The Complete Caves and a view of the hills, p 163 (1977)

TRAT'S CRACK   Cave
ST 4819 5820   L 17m VR 7m Alt 165m
Also known as Trat's Hole, Trat's Rift and Rift Cave, this impressive rift, which is located on the north side of the combe and 10m above road level, was originally choked with scree and silts. Excavated by UBSS during 1921-22, it leads steeply down to a pool of water, which was dived to a depth of 4m by SBSS in 1987 and CDG in 1995. The floor consists of soft silt and no way on was found.

Ref: N. Barrington, W. Stanton, The Complete Caves and a view of the hills, p 163 (1977)

FOXES HOLE   Cave
ST 4823 5822   L 40m VR 10m Alt 174m
Also known as Plumley's Den, Fox's Hole and Bat's Lair, this twin-entrance cave (both with shotholes), yields a low wide chamber where Boyd Dawkins excavated a small Pleistocene deposit in 1860. A short, steep passage (with artificial concrete steps) descends to a second chamber containing two deep pools and the remains of bunks and shelves installed by a covert Auxiliary Unit of the Blagdon Home Guard during the early years of WWII. The cave has a legendary connection with John Plumley, a supporter of the Duke of Monmouth during the famous rebellion. Fleeing from the defeat at the Battle of Sedgemoor in 1685, Plumley was said to have hidden in the cave before eventually being caught and hanged near Locking Manor. However recent research has shown that it was a William, not John, Plumley who was caught up in the Monmouth Rebellion, and he was taken from his sick bed to Ilchester Gaol, where he later died. There is another variation which suggests that 'Jack Plumley' may have been a local highwayman, but this is also likely to have been confused with other accounts, such as the one surrounding Thomas Pocock, a legendary figure who operated from a 'cave' in the Polden Hills during the early 1700s.

Ref: Mendip Underground, MCRA p 143 (2013)

LOST CAVE OF BURRINGTON 3   Lost Cave
A cavern described by William Boyd Dawkins in 1864, which H. S. Hawkins, writing in 1948, claimed to be an 'unknown' site. Unfortunately Hawkins unknowingly confused Plumley's Hole with Plumley's Den, not realising that the latter was another name for Foxes Hole. This subsequently turned out to be the mysterious missing cave.

Ref: D. J. Irwin, The Lost Caves of Mendip, BEC Belfry Bull 505 pp 31-46 (1999)

ROAD ARCH
Cave
ST 4835 5819
L 1m Alt 167m

This large choked arch is located at the roadside on the north bank roughly 100m up-combe from Foxes Hole. There is a tiny cavity in one corner that carries a faint draught.

TOAD'S HOLE
Cave
ST 4834 5806
L 8m VR 8m Alt 198m

Also known as Swallet A, this impressive shaft is located near the top of the cliffs on the north bank of the combe, almost directly above Road Arch. The shaft, which has been mined, is steeply-inclined and is choked at the bottom with leaf mould and boulders. It is almost certainly the site visited by Savory in July 1911 which he mistakenly referred to as Foxes Hole (see Lost Cave of Burrington 5).

Ref: N. Barrington, W. Stanton, The Complete Caves and a view of the hills, p 162 (1977)

LOST CAVE OF BURRINGTON 5
Lost Cave

A cave recorded by J. Harry Savory in 1911. *'It is three quarters way up cliff opposite Ellick Wood just above S curve above E. Brooklet. It shows a bush of yew and some bare rocks from the road. After zigzagging up to it over loose surface scree we found it to be a vertical drop slightly inclining in to the cliff, 5-6 ft diameter all the way down, silted up at the bottom, resembles Plumley's Den but larger, 30 ft deep shown by reflected sunlight, shows promise of further galleries from one or two recesses now choked, a little work might clear these ...'*. Although Lizard Hole has been put forward as a possible contender, the description matches Toad's Hole perfectly.

Ref: D. J. Irwin, The Lost Caves of Mendip, BEC Belfry Bull 505 pp 31-46 (1999)

FROG'S HOLE
Cave
ST 4836 5822
L 3m VR 2m Alt 190m

Located in the same cliffs as Toad's Hole, but slightly lower down and 20m further east, this small cave, which is partially concealed behind a tree, consists of a small chamber formed in a vein of spar with two entrances.

Ref: N. Barrington, W. Stanton, The Complete Caves and a view of the hills, p 81 (1977)

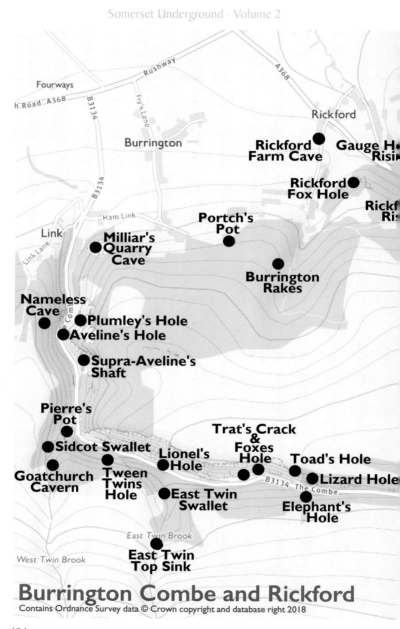

Burrington · Burrington Combe

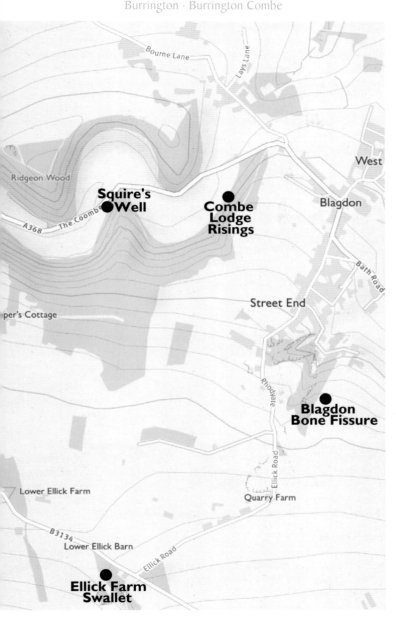

BURRINGTON HAM MINE
Mine
ST 4837 5831 L 8m VR 3m Alt 207m

This small mine was first recorded by WCC in 1983, who entered a small steeply-descending mined passage excavated along a quartz vein. Situated above the combe, on the broad grassy plateau known as Burrington Ham, it was entered via a small triangular hole located in a pit positioned roughly midway along an east-west trending rake which ran alongside a major footpath parallel to the combe. This is almost certainly the same site christened Burrington Ham Top Dig by another WCC team in 1989, who extended the dig into a small mud-choked chamber before abandoning the site and filling it back in. The area has been largely levelled, and the site is now indistguishable from several other nearby mined pits, although the top rung of an old wooden ladder has been spied occasionally poking up through the bracken and leaf mould.

Ref: P. G. Hendy, Caving Diary Vol.8 pp 1199-1200 (1983) & From the Log, WCC Jnl 20(221) p 61 (1989)

BOULDER SHAFT
Cave
ST 4835 5814 L 6m VR 6m Alt 170m

Located on the south bank, roughly 10m above the road and almost directly opposite Road Arch, this vertical hole, which is probably an old mineshaft, is the only one still open of several similar shafts situated at this level on this side of the combe. One, 65m to the west contains a tiny cavity (**E Hole**, ST 4829 5816, L 1m), while 20m further up the combe, there is another tiny hole covered with a large boulder at the foot of a tree with twin trunks which offers a glimpse into a slightly larger cavity.

Ref: N. Barrington, W. Stanton, The Complete Caves and a view of the hills, p 44 (1977)

ELEPHANT'S HOLE
Mine
ST 4842 5814 L 17m VR 7m Alt 174m

Also known as Spider Hole, Swallet C, Calamine Hole, Pig Hole and Pig's Hole, this mine, which is clearly visible from the road, is located on the south bank some 80m up-combe from Road Arch. It consists of a roomy adit which leads to a steeply-descending slippery rift with a tight upward connection to Elephant's Rift located on the hillside directly above. It was first dug by the UBSS in 1919-21 and has been looked at by various clubs at intervals since.

Ref: N. Barrington, W. Stanton, The Complete Caves and a view of the hills, p 70 (1977)

ELEPHANT'S RIFT
Cave
ST 4842 5814 L 12m VR 10m Alt 182m

Also known as Elephant Rift, this long narrow open gash on the hillside yields a steeply-descending bedding plane which connects directly into Elephant's Hole below the adit.

Ref: N. Barrington, W. Stanton, The Complete Caves and a view of the hills, p 70 (1977)

LIZARD HOLE
Cave
ST 4844 5819 L 10m VR 9m Alt 195m

Also known as Swallet B, this steep mined shaft is located 20m above the road on the north side of the combe, and almost directly opposite Elephant's Hole. Along with Toad's Hole, it was put forward as a possible contender for the cave visited by Savory in 1911 (see Lost Cave of Burrington 5).

Ref: N. Barrington, W. Stanton, The Complete Caves and a view of the hills, p 108 (1977)

PIPSQUEAK HOLE
Collapse
ST 4834 5807 L 4m VR 4m Alt 198m

This small collapse is located high up on the south flank of the hill on the edge of the wood almost directly above Boulder Shaft. It appeared in 1969, and was dug by Yeo Trogs Group in 1970, but revealed only earth mixed with sandstone pebbles. Largely infilled, its position on the boundary of the field and the wood suggests the presence of a broken field-drain.

Ref: N. Barrington, W. Stanton, The Complete Caves and a view of the hills, p 122 (1977)

HAWKESWELL QUOIT
Spring
ST 4839 5794 Alt 217m

Located at the top south-east corner of the field immediately above Pipsqueak Hole, this small walled pit, surrounded by broken rock, may once have enclosed a spring. The unusual name (which is derived from the name of the field) suggests that a megalith may have stood nearby, although no evidence for its existence has been presented.

Ref: P. Quinn, The Holy Wells of Bath and Bristol Region, Logaston: Almeley, p 164 (1999)

ELLICK FARM SWALLET
Sink
ST 4927 5780 Alt 216m

This marshy sink was dug inconclusively by BEC c.1956. The stream, which may be the one seen in the lower reaches of Lionel's Hole, rises on the north-east flank of Blackdown (**Ellick Springs**, ST 4922 5731) before sinking in a large wet area of the valley floor. It has been traced to both Rickford and Langford Risings.

Ref: N. Barrington, W. Stanton, The Complete Caves and a view of the hills, p 70 (1977)

MOLE HOLE
Mine
ST 4927 5818 L 20m VR 10m Alt 210m

In 2007, a tractor wheel sinking into an otherwise flat field revealed this small ochre working. WCC descended a loose 10m shaft which yielded a narrow calcite vein, roofed with boulders. It has been filled in.

Ref: P. G. Hendy, A Sunday Excursion – The Mole Hole. Burrington, WCC Jnl 29 (306) pp 78-79 (2007)

WEST TWIN BROOK VALLEY

PIERRE'S POT
Cave
ST 4763 5837
L 940m VR 51m Alt 129m

Located just off the main path on the east bank of West Twin Brook, this sporting cave comprises a small but complex upper series, and an extensive and well-decorated lower series, much of which is only accessible through sumps and difficult squeezes. A once passable connection with Pseudo Nash's Hole is now blocked. The main streamway in the lower series originates from East Twin Swallet with a further inlet (Western Inlet) fed by Flange Swallet and Goatchurch Cavern. The water can be pursued upstream through two sumps until it becomes choked with silt. An extensive series of small fossil passages before Sump 1 leads to the well decorated Hanging Gardens. Downstream, two sumps lead to another extensive series of fossil passages containing some superb formations. The passage ends in Downstream Sump 5, a partly-choked horizontal phreatic tunnel, trending towards Langford Rising. The cave was discovered by WCC after a brief excavation during 1983, and subsequent digging yielded part of a bear skull. The lower streamway was reached in 1987 and the upstream sumps pushed to their current conclusion shortly after. The downstream sumps were passed by CDG in 2011. 'Pierre' was another nickname of Mike 'Fish' Jeanmaire, a colourful character and long-standing chairman of the CDG, who sadly passed away in 2010.

Ref: Mendip Underground, MCRA pp 234-239 (2013)

PSEUDO NASH'S HOLE
Cave
ST 4761 5835
L 8m VR 4m Alt 135m

Located adjacent to the path on the east bank of West Twin Brook, this cave comprises a low bedding plane which descends steeply to a choke. The cave once connected to Pierre's Pot, entering just below the tight vertical squeeze leading to the lower series. The connection was dangerous and collapsed c.1987.

Ref: N. Barrington, W. Stanton, The Complete Caves and a view of the hills, p 125 (1977)

JOHNNY NASH'S HOLE
Cave
ST 4761 5835
L 6m VR 5m Alt 141m

Named after the UBSS member who dug it in 1949, this cave comprises a very tight, steeply-descending phreatic tube, which terminates in a tiny chamber. It lies at the foot of a small cliff located directly above Pseudo Nash's Hole.

Ref: N. Barrington, W. Stanton, The Complete Caves and a view of the hills, p 102 (1977)

Pierre's Pot. Photo by Clive Westlake

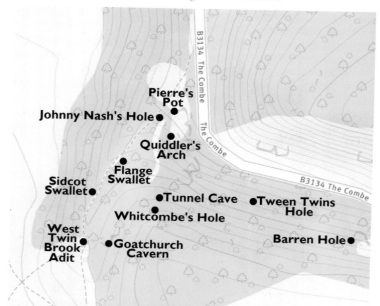

West Twin Brook Valley

Contains Ordnance Survey data © Crown copyright and database right 2018

QUIDDLER'S ARCH
Cave
ST 4762 5834
L 2m VR 1m Alt 154m

Located in a rocky spur high up on the east bank of the West Twin Brook, this insignificant hole consists of nothing more than a small phreatic arch leading to an impenetrable second entrance. Stanton apparently assigned it this 'trifling' name because they lacked a 'Q' in 'Complete Caves!'

Ref: N. Barrington, W. Stanton, The Complete Caves and a view of the hills, p 125 (1977)

FLANGE SWALLET
Sink
ST 4758 5832
L 10m VR 10m Alt 140m

The West Twin Brook sinks against a cliff face inside a fenced enclosure. Also known as Peter Bird's Dig, Swallet H, West Twin Brook Swallet and Minnie Ha Ha, the swallet has been dug by several groups, including the UBSS, who in 1959-64 excavated a wide shaft to reach a small choked joint. The shaft refilled during the 1968 floods. The stream has been MKHP spore-tested to Langford Rising (17 hours) and Rickford Rising (19 hours), but surprisingly, the main flow of water has not been detected anywhere inside nearby Pierre's Pot. A moderate positive test was achieved between the sink and Pierre's Western Inlet in 1987, but the size of the respective streams differs so markedly that the test probably only represents minor seepage from the surface.

Ref: N. Barrington, W. Stanton, The Complete Caves and a view of the hills, p 78 (1977)

SIDCOT SWALLET
Cave
ST 4755 5829   L 200m VR 28m Alt 149m

This cave, located above the stream on the west bank of West Twin Brook, provides a variety of squeezes, crawls, sporting climbs and small chambers, and is particularly popular with novice parties. Many calcite formations remain intact in the farther reaches. The entrance was first noted by Boyd Dawkins in the 1860s but wasn't dug open until 1925 when SSSS finally gained entry.

Ref: Mendip Underground, MCRA pp 318-320 (2013)

YEW TREE SWALLET
Sink
ST 4757 5828   L 3m VR 3m Alt 147m

Also known inaccurately as Hawthorn Tree Swallet, and also as Kong's Hole, only the yew tree now remains of the original sink for the West Twin Brook. It was dug inconclusively by SSSS during the 1920s and buried beneath tipped rubble extracted from West Twin Brook Adit c.1939. In 1962, UBSS dye-tested the water to the Water Chamber of Goatchurch Cavern, and from there to Langford and Rickford Risings. Old maps indicate that the stream once sank close to the junction with the main combe.

Ref: N. Barrington, W. Stanton, The Complete Caves and a view of the hills, p 182 (1977)

WEST TWIN BROOK SPRING
Spring
ST 4755 5824   Alt 156m

A small spring housed inside a stone tank.

WEST TWIN BROOK ADIT
Tunnel
ST 4755 5820   L 300m VR 3m Alt 158m

Also known as Burrington Adit, the Waterworks Tunnel and Burrington Waterworks Adit, this tunnel was driven by Axbridge Rural District Council (1941-44) in an attempt to tap into the vast reservoir of water believed to exist within the Old Red Sandstone cap of Blackdown. The venture was only a partial success and the supply was abandoned in 1968 when the pipe work was damaged during the great Mendip flood. The adit remains open but is significantly waterlogged.

Ref: N. Barrington, W. Stanton, The Complete Caves and a view of the hills, p 172 (1977)

Goatchurch Cavern. Photo by Steve Sharp

GOATCHURCH CAVERN
Cave
ST 4758 5822 L 1300m VR 59m Alt 162m

This twin-entrance cave has had many names including Burrington Combe Cavern, Burrington Cavern, Goat's Hole, Goechurch Cave and possibly Guy Hole. It is among the most popular on Mendip, and is a particular favourite with novices. Its complex network of predominantly dry passages provides a number of tight squeezes, short climbs and gymnastic manoeuvres, and many of the features carry familiar and evocative names, such as the Giant's Stairs, the Coffin Lid and the Drainpipe. The small stream, which emerges from a tight fissure at the upper end of Water Chamber, emanates from nearby Yew Tree Swallet and has been dye-tested to both Langford and Rickford Risings, just over 1.5km to the north-west and north-east respectively. Writing in 1829, John Rutter described Goatchurch as '*an extensive and intricate cavern, but little known. Its entrance, on the side of the hill, is small, but on advancing, it is found more spacious, and presents magnificent masses of stalagmite. A second descent of about 6m, leads to another portion of the cavern, which opens into numerous ramifications, so intricate that even the miners, who reside in the vicinity, find it requisite to use twine as a guide for their return; the termination of these passages have never yet been thoroughly explored. A stream of water runs across the floor of one part of the cavern, forming a waterfall at a great distance from the entrance...*' Early records of Goatchurch Cavern include the archaeological work of William Beard (1830), William Boyd Dawkins (1860) and UBSS (1923-25), during which time the Tradesman's Entrance was opened. Material relating to the Pleistocene period, including the bones of mammoth, horse, bear, hyaena and cave lion, was recovered. There is at least one old shothole present in the cave (near the Coffin Lid), and the remains of steps and iron railings in the main entrance passage are all that is left of an attempt, by a previous owner, James Gibson, to establish a show cave c.1901-04. There is another impassable fissure situated between the two entrances which probably connects with The Maze.

Ref: Mendip Underground, MCRA pp 151-155 (2013)

GUY HOLE
Lost Cave

An apparently lost cave mentioned in an unpublished manuscript written by the antiquary, John Strachey in 1736 and housed in the Taunton county archives. Known to him since at least 1720, he described it as '*... under this fortification is an hole or Cave called Guy Hole, altogether as remarkable as that at Woky but the former being near a City & this remote from any place of Entertainment is not often visited by Travellers ..*'. The cave purportedly lay somewhere below the fortifications at Dolebury near Burrington Combe. Investigations by Irwin suggest that the most likely explanation is that it was an alternative name for Goatchurch Cavern.

Ref: D. J. Irwin, The Lost Caves of Mendip, BEC Belfry Bull 505 pp 31-46 (1999)

EAST TWIN BROOK VALLEY

SPAR POT   Cave
ST 4796 5818   L 183m VR 34m Alt 150m
Also known as Clive North's Dig, no sign now remains of the original entrance to this complex of small fossil passages, which was washed open during the floods of 1968. The cave was dug open by ACG during 1970-71 but closed shortly after at the landowner's insistence. Happily, SBSS recovered the cave in 1983, by engineering a connection from East Twin Swallet, which joined the cave close to the original entry point. The cave is now referred to as the Spar Pot Series of East Twin Swallet, and its length and depth details have been incorporated into the latter (see below).

Ref: N. Barrington, W. Stanton, The Complete Caves and a view of the hills, p 146 (1977)

EAST TWIN SWALLET   Cave
ST 4795 5815   L 407m VR 40m Alt 155m
This cave, excavated by the UBSS in 1935, is the main sink point for the East Twin Brook, and consists of a steeply-descending stream passage which passes through a series of chambers. These were once roomy but have since been largely filled with digging spoil. In 1983, SBSS engineered a connection with Spar Pot (now the Spar Pot Series), an interesting complex of passages which had been explored by ACG in 1970-71 from a separate entrance and sealed by the landowner a year later. In 1968 MKHP spore-tested the stream to both Langford Rising (17 hours) and Rickford Rising (6 hours), 1.8 km to the north-west and 1.3 km to the north-east respectively. The water passes through Pierre's Pot en route. Work by SBSS continues.

Ref: Mendip Underground, MCRA pp 114-115 (2013)

EAST TWIN MIDDLE SINK   Cave
ST 4793 5802   L 6m VR 3m Alt 174m
Also known as Dreadnaught Holes, this site comprises three small phreatic holes, located on either side of the East Twin Brook. They were dug open by Dreadnaught Caving Club c.1971.

Ref: N. Barrington, W. Stanton, The Complete Caves and a view of the hills, p 65 (1977)

EAST TWIN TOP SINK   Cave
ST 4792 5796   L 6m VR 1m Alt 185m
The only visible sign today of this sporadic dig site is a small pool of water held back by a dam of rocks located a short distance upstream of the Limestone Link, the main path that crosses the valley floor. In wet weather, a large stream sinks here, and the sink point has been attacked by several clubs, but so far only a few unstable sculpted holes have been revealed.

Ref: N. Barrington, W. Stanton, The Complete Caves and a view of the hills, p 65 (1977)

RICKFORD

Rickford Rising is the larger of the two risings fed by the Burrington streams and emerges in the lower reaches of Rickford Combe, a deeply-incised valley carved through Dolomitic Conglomerate, which may be an overflow channel resulting from an ice-dammed lake. The stream enters an idyllic mill pond, which is largely the creation of the Wills family in the late 19th century, who also constructed a boat house and the colourful Baptist chapel. This striking building operated until the 1960s, and is now a Masonic lodge. The water eventually joins the Congresbury Yeo, en route to the Bristol Channel.

PORCH'S POT   Cave
ST 4815 5902   L 12m VR 6m Alt 115m

Also recorded as Porch's Pot, this short phreatic passage was uncovered in a garden in 1955 and permanently closed soon after. The exact location is uncertain.

Ref: N. Barrington, W. Stanton, The Complete Caves and a view of the hills, p 124 (1977)

BURRINGTON RAKES   Mines
ST 4829 5896   L 1m VR 1m Alt 131m

Numerous calamine rakes can be found on the wooded slopes to the south of Burrington village. The most impressive, excavated in Dolomitic Conglomerate, lie on the western edge of the valley overlooking The Hill Gardens, a series of historic allotments created for local labourers c.1830. Now thoroughly overgrown, the 'gardens' later provided cover for the construction of the Bailey Bridges used by Allied soldiers during the D-Day landings. In 1919, Savory recorded east-west rakes running all the way up from The Hill Gardens as far as Ellick Farm.

RICKFORD FOX HOLE   Cave
ST 4860 5920   L 6m VR 1m Alt 88m

Located near the top of an outcrop of Dolomitic Conglomerate, this low tunnel ends in a tiny circular chamber. It was dug by local youths c.1938. There are several overhangs and impenetrable solutional rifts in the surrounding cliffs. Collectively known as **Orchard Holes**, they were first officially recorded by Stanton in 1945.

Ref: W. I. Stanton, Logbook 2 p 89 (1945) & N. Barrington, W. Stanton, The Complete Caves and a view of the hills, p 130 (1977)

RICKFORD FARM CAVE   Cave
ST 4847 5936   L 53m VR 8m Alt 67m

Also known as Wyatt's Cave, this cave was uncovered in 1959, by workmen digging the foundations for a new cowshed. A narrow 7m entrance pitch yielded almost 50m of high, narrow and locally well-decorated rift, all in Dolomitic Conglomerate. The cave was dug on several occasions, but the entrance appears to have been sealed when the farm's outbuildings were converted to private dwellings during the 1980s. The entrance probably lies directly beneath a raised rectangular patch of gravel, situated in the front garden at the southern end of the new house.

Ref: N. Barrington, W. Stanton, The Complete Caves and a view of the hills, p 130 (1977)

BATCH WELL

Well

ST 4853 5949

Alt 50m

This spring-fed roadside well, is located halfway up the lane known as The Batch. There are other springs in the gardens nearby.

GAUGE HOUSE RISING

Rising

ST 4873 5926

Alt 60m

This small well is located in an artificial chamber situated immediately below the waterfall at the downstream end of Rickford Pond. It seldom runs dry, and a small channel carries any overflow back to the main flow. Some of the water probably passes underground in pipes to emerge further down-valley from a roadside pump situated opposite the Gauge House. This was built in 1895 by the Bristol Waterworks Company, and houses both the regulating weirs used for controlling the flow of waters, and the underground pipe which feeds Blagdon Lake.

Ref: N. Barrington, W. Stanton, The Complete Caves and a view of the hills, p 130 (1977)

RICKFORD RISING

Rising

ST 4879 5916

L 5m VR 5m Alt 61m

Situated across the road from Rickford's idyllic pool, this, the larger of the two Burrington catchment risings, is largely fed by percolation water (90-95%) and is an important source of water for the Bristol Waterworks reservoir of Blagdon Lake. The rising is probably fed by all the Burrington caves, but proven links include Read's Cavern, Flange Swallet, Pierre's Pot (sourced from East Twin Swallet, Goatchurch and Flange), Yew Tree Swallet (via Goatchurch Cavern), East Twin Swallet and Ellick Farm Swallet, along with two more distant sources at Ubley Hill Pot and Lamb Leer Cavern. The results are based on several different tracing attempts (not all successful), and the wide variations in recorded times render them almost meaningless. The rising responds rapidly to heavy rain and dwindles to almost nothing in times of drought, and it has been postulated that the relatively rapid response times of some of the traces may be due to the water travelling through very small tubes, the so-called 'hosepipe hypothesis'. A scaffolded square-sided vertical shaft in Carboniferous limestone and Dolomitic Conglomerate, has been pursued by CDG divers for 5m. The rising may represent a capture from an ancient resurgence at Aveline's Hole and may in turn be being captured by Langford Rising. Non-cavers sometimes refer to it as Shirborne Spring, an Old English term meaning 'radiant stream'.

Ref: N. Barrington, W. Stanton, The Complete Caves and a view of the hills, p 130 (1977) & A. R. Farrant, G. J. Mullan & A. A. D. Moody, Speleogenesis and Landscape Development in the Burrington Area, Somerset, UBSS Proc 24 (3) pp 207-252 (2008)

SQUIRE'S WELL
ST 4928 5908

Spring
Alt 70m

Also known as The Squire's Well, this rising is situated in an unimpressive pit of Dolomitic Conglomerate adjacent to the road. It only discharges in very wet conditions, possibly due to 'spring sapping' in the lower reaches of Rickford Combe. It may represent a short-lived capture from an ancient resurgence at Aveline's Hole, which was subsequently captured, in turn, by Rickford Rising. Once considered 'worthy of investigation', the spring is almost completely choked with stones, and any water that does emerge flows immediately into a solidly constructed road drain. The spring lies at the lowest point of Fuller's Hay, a large expanse of private woodland which contains some minor signs of iron mining activity. This wood was suggested as a possible location for the 'lost' Blagdon Bone Fissure, but recent investigations carried out by the author strongly suggest otherwise.

Ref: N. Barrington, W. Stanton, The Complete Caves and a view of the hills, p 148 (1977)

COMBE LODGE RISINGS
ST 4968 5913

Springs
Alt 91m

After heavy rain, clear water rises from at least eight small intermittent springs located on the boundary of the wood and the eastern edge of 'The Park', the name of the field located directly opposite the driveway to Combe Lodge.

Rickford Pond. Photo by Keith Savory

Somerset Underground · Volume 2

NORTH MENDIP

Burrington Combe may be the best known of the combes and gorges which punctuate Mendip's steep northern escarpment, but it is by no means alone. Compton Combe, Harptree Combe and Burges's Combe are no less impressive, and all offer subterranean excursions of interest. Compton Combe was noted for its 20th century ochre production, echoing the 18th century calamine workings in Harptree Combe. Both were pre-dated by lead production which was worked in many places along the northern flank, but most notably at Eaker Hill, Buddle's Wood and on Smitham Hill, where Mendip's sole surviving complete chimney stands sentinel. Natural caves are less plentiful, but are often unusual in nature, particularly those near Greendown and Chewton Mendip, some of which appear to be hydrothermal in origin. The largest cave by far however remains Lamb Leer, which for 80 years held the unexpected distinction of being the world's deepest explored cave!

Smitham Chimney. Photo by Rob Taviner

BLAGDON

Derived from Old English bloec and dun (meaning black or bleak down), Blagdon contains two important springs, but from a subterranean point of view is probably most famous for a lost bone fissure, discovered by iron miners c.1872.

LOWER WELL   Spring
ST 5026 5892   Alt 88m

This spring rises in a triangular roadside enclosure, close to the parish church of St Andrew, and despite its proximity to the graveyard, was once used as a village water supply. Rev. Augustus Toplady, a staunch Calvinist and writer of the hymn Rock of Ages, was a curate of the church from 1762 until he became a priest in 1764. He allegedly penned the hymn on the back of a playing card during a storm in Burrington Combe in 1763, although this claim is now subject to considerable doubt.

Ref: P. Quinn, The Holy Wells of Bath and Bristol Region, Logaston: Almeley, p 163 (1999)

TIMS WELL   Spring
ST 5026 5892   Alt 92m

Originally the main village water supply, this spring, also known as Timsell Well, rises in a small pond contained within a walled enclosure. It used to supply baptismal water to the village church.

Ref: P. Quinn, The Holy Wells of Bath and Bristol Region, Logaston: Almeley, p 163 (1999)

STREET END HOLE   Cave
ST 4992 5841   L 1m VR 1m Alt 175m

This miniscule hole, which is little more than a solutionally-enlarged joint, is situated beneath a low cliff of Dolomitic Conglomerate, in a small side offshoot of the narrow defile which connects Street End Land with Ellick Road. There are a number of equally impenetrable holes nearby, and yet more in an interesting quarried hollow a short distance further north-west.

BLAGDON BONE FISSURE   Lost Cave
ST 4999 5836   L 13m VR 13m Alt 192m

In 1872, miners searching for iron ore '*near the top of Blackdown Ridge above Blagdon*', uncovered a fissure 13m down, containing bones of bos and rhinoceros. Possibly part of a larger cave, it was filled in before it could be properly investigated. Apart from its position near the top of the ridge, the only other clue to its whereabouts was that it was in Dolomitic Conglomerate, and that there was a shallow valley located 50 yards to the west of the mine. E. K. Tratman was the first to search for it, but was unsuccessful, and a number of possible locations have been suggested subsequently, including Swancombe and Merecombe Woods towards Ubley, and Fuller's Hay, towards Rickford. However investigations by the author during the course of creating this guide, clearly point towards this filled shaft as being the most likely candidate and is presumably the same one worked by T. S. Smith of Bristol Mineral Co. during the 1870s.

Ref: Geological Section, Proceedings of the Bristol Naturalists' Society, Vol 1 pp 137-138 (1874-6), M. Clarke, N. Gregory and A. Gray, Earth Colours – Mendip and Bristol Ochre Mining, MCRA, p 129 (2012) & R. M. Taviner, Blagdon Bone Fissure, available online via www.mcra.org.uk (2017)

SWANCOMBE WOOD HOLE — Cave
ST 5083 5826   L 1m VR 1m Alt 170m

Situated a short distance down-valley of a large limekiln and above a prominent outcrop of Dolomitic Conglomerate, this site consists of three largely-impenetrable small rifts surrounding a calcite vein. There are similar, but even smaller features nearby, and despite the apparent lack of mining evidence, several small trial workings (probably for iron) were reputably backfilled. One of these workings may have been the site visited by BEC in 1945/46. It was called **Swancombe Hollow Dig** and a survey was made, although this has since been lost. Some commentators have suggested that this area might be a good location for the fabled Blagdon Bone Fissure. However, it lies several hundred metres east of the village and the presence of a much better, and closer, candidate, strongly suggests otherwise.

Ref: N. Barrington, W. Stanton, The Complete Caves and a view of the hills, p 152 (1977

The probable location of Blagdon Bone Fissure. Photo by Rob Taviner

UBLEY & COMPTON MARTIN

The main speleological interest on this portion of the hill is confined to the various iron ore (red ochre) mines, which began life as small opencast workings in the 19th century. The iron oxide, which lies in Dolomitic Conglomerate, dates to the Triassic, and was probably deposited underwater as a layered mass in a desert environment fed by hot springs. Much of this work was centred on Compton Combe, and in particular the Redding Works Quarry, where underground operations started in the 1920s, and continued extracting pigment for use in paint manufacture right through until 1957. Compton Martin was the birthplace of St Wulfric (1080–1154), a renowned hermit and miracle worker, who successfully predicted the death of Henry I.

UBLEY

UBLEY CONDUIT   Conduit
ST 5212 5855   Alt 50m
Uncovered in 1979 by workmen laying a pipe for Bristol Waterworks, this unusual site is a tiny subterranean streamway formed in Keuper Marl located 2m below the surface. The NGR given is for the resurgence although a number of associated subsidence collapses have appeared on the hillside above. They have been backfilled with rubble.

Ref: W. I. Stanton, Ubley Conduit – An underground streamway in Triassic clay at Ubley, North Mendip, UBSS Proc 16 (2) pp 105-108 (1982)

ROOKERY FARM SPRING   Spring
ST 5213 5821   Alt 63m
A small spring which is probably piped from the fields on the opposite side of the main road.

MERECOMBE WOOD SHAFT   Mine
ST 5157 5790   L 8m VR 8m Alt 185m
This open mineshaft, which is effectively sealed beneath a metal grill and a large boulder, possibly hasn't been descended since the miners departed. Surrounded by a collar of debris, the shaft descends steeply, and there is evidence for at least two other similar shafts in the undergrowth nearby. The shafts lie near the head of a wooded valley, where there are lots of open bedding planes and natural fissures visible. These are in Dolomitic Conglomerate, although none appear penetrable. The location of the shafts, together with evidence of mining in the field to the east, have led some

commentators to suggest that this area might be a good location for Blagdon Bone Fissure. However, the shafts lie some distance south of the favourable iron-rich unconformity separating the limestone from the Dolomitic Conglomerate, and the presence of a much better candidate further west, coupled with its significant horizontal distance from the village (approx. 1.5 km), suggests otherwise.

HAZEL MANOR MINE Mine
ST 5341 5690 Alt 248m

Little surface evidence now remains of this small iron mine, which was sunk by Mr. W. Thomas in 1867. He operated a similar venture at Wigmore Farm (see All Eights Mineshaft) during the same period, and later went on to work pits at Lamb Bottom. There is also a well nearby which may have started life as a trial shaft. It is clearly marked on old maps (ST 5309 5696). Erected during the Elizabethan period, Hazel Manor was one of only two manor houses built on the Mendip plateau. Sadly it burned down in 1929.

Ref: J.W. Gough, The Mines of Mendip, David & Charles p 247 (1967)

UBLEY HILL FARM RIFT Mine
ST 5197 5767 L 17m VR 9m Alt 225m

The Rev. William Jones, a vicar with a penchant for bone caves, recorded a site in a field at Ubley Hill Farm, where '... *a stone dropped into the hole may be heard for several seconds in its' downwards course* ....' This may be the site of a small shaft explored by the MCG in 1984 that appeared following a collapse in a filled-in rift. The hole, which has since been capped and obscured, now lies in a scrambling course, close to its northern edge.

Ref: A.J. Knibbs, Ubley Hill Farm Rift, MCG News No.174 pp 8-9 Dec (1984) & D.J. Irwin, The Lost Caves of Mendip, BEC Belfry Bull 505 pp 31-46 (1999)

COMPTON MARTIN

COMPTON MARTIN RIFT CAVE Lost Cave

In 1921, twelve UBSS members visited a quarry owned by a Mr. Bath at Compton Martin, to examine a number of holes, one if which emitted the sound of a running stream. UBSS apparently worked at this site for the next five years, but unfortunately the logbook covering this period was one of those destroyed in the Bristol blitz during WWII and both the exact location and any results they might have achieved, are lost to us. However, the more northerly quarry had yet to be developed, so the holes can only have been located in either the central Cliff Quarry, or in the vicinity of Compton Martin Ochre Mine, and the fact that underground commercial operations appear to have begun at exactly the same time that UBSS investigations ceased (1926), strongly suggests that the mine simply developed from the earlier work.

Ref: D.J. Irwin, The Lost Caves of Mendip, BEC Belfry Bull 505 pp 31-46 (1999)

CLIFF QUARRY HOLE   Cave
ST 5409 5672   L 2m VR 2m Alt 152m
This waterworn hollow, formed inside a calcite vein, may be a remnant of the lost holes visited by UBSS in 1921, although Compton Martin Ochre Mine seems a more likely alternative. There are some excellent dog-tooth crystals, and the hole is surrounded by numerous solutional features and active tufaceous cascades.

CLIFF QUARRY RIFT   Cave
ST 5409 5670   L 3m VR 3m Alt 152m
Located a few metres along the same quarry face as Cliff Quarry Hole, this rectangular gash may also be a remnant of the lost holes visited by UBSS in 1921. Dog-tooth crystals are evident throughout, particularly in several large detached blocks of calcite embedded in the fill.

COMPTON MARTIN OCHRE MINE   Mine
ST 5420 5670   L 400m VR 18m Alt 152m
Also known as The Redding Works, these maze-like workings follow a slightly-inclined bed of ferruginous conglomerate from which red ochre was extracted using pillar and stall techniques. Several large geodes filled with coarsely-crystalline calcite are visible, and interesting forms of fungi can also be found. During a major renovation project in 2007, a number of very rotten pit props were replaced and areas of fallen 'deads' were neatly restacked. The majority of the mine is now easy walking and at one point rails may be seen in the floor. Surface activity probably dates from the late 19th century, but underground extraction of red ochre began around 1926 and remained in operation as recently as 1957. The commercial operations may well have been a continuation of investigations carried out by UBSS in 1921-26. The ochre was chiefly used as a pigment in paint manufacture. It is an SSSI, a Geological Review Site and thanks to the presence of hibernating greater horseshoe bats, a European designated Special Area of Conservation (SAC). The earliest documented visit by cavers was in 1974, when it was located and explored by SMCC. There were probably several other entrances into this mine, evidence of which can be seen in either direction along the cliff. All have since collapsed, or remain impenetrable.

Ref: Mendip Underground, MCRA pp 98-99 (2013)

COMPTON COMBE SHELTER   Cave
ST 5390 5649   L 3m VR 3m Alt 200m
Located in a narrow, gorge-like section of the combe, this obvious overhang contains a narrow rift formed in a calcite band.

COMPTON COMBE MINESHAFT   Mine
ST 5370 5629   L 1m VR 1m Alt 228m
Located in an extensive area of overgrown quarry workings, directly opposite a truly horrible car dump, this is a choked mineshaft, partially covered with a metal grill.

RAG SPRING
ST 5470 5659

Spring / Holy Well
Alt 170m

Also recorded as Compton Hill Spring, this recently restored, ancient quasi-religious dipping well was once the only source of water for the village. It has an arched well head, reached by a flight of crude steps. The name is a reference to an ancient practice whereby people would hang the rags they'd used to wash afflictions with in the surrounding trees in the belief that as the rags gradually wasted away, they would take the affliction with it. The only recorded one of its kind in Somerset, the water was said to be good for both the eyes and sprained ankles. The spring is the largest of several which flow from the Dolomitic Conglomerate, and unite to form a stream feeding the village pond, where a well of an entirely different type can be seen. This is the **Dipping Well** (ST 5450 5703), which is housed inside a '*blue tardis*', alongside the village pond. The hand operated pump still functions, but the water, sadly, is no longer fit for human consumption. The Compton Martin springs have no proved feeders although oil contamination was reported following the illegal dumping of thousands of gallons of oil into the Devil's Punch Bowl in 1967.

Ref: P. Quinn, The Holy Wells of Bath and Bristol Region, Logaston: Almeley, p 174 (1999)

Rag Spring. Photo by Rob Taviner

BARROW WELL
ST 5374 5734

Spring
Alt 75m

This rather dilapidated spring derives its name from a prehistoric barrow, all traces of which have sadly now been lost. Local folklore states that the well was haunted by the ghost of a woman washing 'cabbages!' Bizarre though that sounds, both cabbages and natural spring water have long been associated with fertility.

Ref: P. Quinn, The Holy Wells of Bath and Bristol Region, Logaston: Almeley, p 173 (1999

EAST HARPTREE

East Harptree is another of those villages whose history rather belies its size. It was the site of Richmont Castle (French for 'fine hill'), an 11th century fortification which was captured by King Stephen during his civil war with Matilda. Later visited by King John, this, the family seat of the de Gourney's, was also home to the Harptree minery court, one of four such institutions which controlled the mineral wealth of Mendip. At its height, the castle boasted a deer park and an ornamental lake, and it survived until c.1540, when it was pulled down by Sir John Newton, partly to build the original Eastwood Manor. (The current house was built in 1871 and was one of the first houses in Somerset to have electric lights). Sir John's truly spectacular tomb can be seen in the porch of East Harptree church. Yellow ochre was probably extracted throughout the parish for centuries, particularly in the fields overlying the Jurassic Harptree Beds and much of the ore was washed in the stream behind Pitt Farm, before being sent on to several factories which were erected in the village to process it. Red ochre was found near Eastwood Manor and around Greendown, and 'Lapis Calaminaris' (calamine) mines were recorded by Collinson in 1791, an industry that continued in fits and starts into the 19th century. It was the lead industry however that was the most important, occupying the high ground of Smitham Hill for centuries before finally culminating in the 'East Harptree Lead Works Co. Ltd', who operated between 1870 and 1880. Work ceased altogether in the 1890s, and today all that survives of this once mighty industry are some overgrown slag heaps, some 'gruffy ground' and the well-preserved lone chimney which once dominated the skyline overlooking the village. 'Smitham' is the name given to small lumps of ore scavenged by free miners to avoid paying lead lot duties.

HARPTREE COMBE

A nature reserve, and an SSSI, the attractive valley of Harptree Combe plays host to a number of interesting features, including a spectacular Grade-II aqueduct, which dates from 1851, and is thought to be the oldest surviving structure of its type in the country. The combe is part limestone, and part Dolomitic Conglomerate, with the latter being of a similar nature to the silicified standing stones at Stanton Drew. This has led one archaeologist to suggest that at least some of the stones may be been quarried in this vicinity and floated down the Molly Brook when water levels were much higher than those seen today. The combe was home to an important calamine (smithsonite) industry and several old workings remain open, some of which may be the 18th century 'Lapis Calaminaris' mines recorded by Collinson.

HARPTREE COMBE MINE 1 — Mine
ST 5620 5567   L 6m VR 2m Alt 135m

Also logged by Phil Hendy as Mine 4001, this short tunnel, located on the west bank of the Combe Lane tributary valley, is an abandoned calamine working. The mine contains a vein of blackened dog tooth spar and some good shotholes. There is further evidence of opencast mining in the ramparts of Richmont Castle, situated directly above. Hendy described the most obvious one as a sloping pit, in 1967.

Ref: A. D. Oldham, The Mines of Harptree Combe, SVCC Ntr 1(2) pp 3-4 (1963) & P. G. Hendy, Mines of East Harptree Coombe, SVCC News 9(3-4) (1967)

HARPTREE COMBE WATERWORKS TUNNEL — Tunnel
ST 5617 5587   L 700m+ VR 30m Alt 130m

A 5m descent through this access hatch (locked), gives access to part of the Line of Works, a 16 km long subterranean tunnel and aqueduct system built in the 1840s to carry water from as far afield as Chewton Mendip to Barrow Reservoirs. In 1974 and 1975, Stanton followed the oval-shaped masonry lined tunnel upstream for several hundred metres, passing through several sections of solid Dolomitic Conglomerate en route. After passing beneath two shafts to the surface (20m and 30m) he eventually reached a point where water entered from an old natural stream channel. Downstream, leads to the steel pipe aqueduct.

Ref: W. I. Stanton, Logbook 13 pp 97-98 & 109 (1974-75)

AQUEDUCT HOLE — Cave
ST 5619 5601   L 2m VR 1m Alt 115m

This tiny waterworn fissure, located on the east bank of the combe close to the lower end of aqueduct, quickly closes down to nothing.

LOWER HARPTREE COMBE FISSURES — Fissures
ST 5626 5608   L 2m VR 1m Alt 110m

The grid reference is centred on the most prominent of several tiny fissures and bedding planes which penetrate the exposures of Dolomitic Conglomerate on both sides of the lower combe.

KING'S CASTLE WELL — Holy Well & Conduit
ST 5612 5587   Alt 115m

Also known as the Harptree Combe Resurgence, and the Harptree Wishing Well, this spring, which debouches a high volume of water in flood conditions, lies near the foot of Richmont Castle and adjacent to the point where the Combe Lane tributary valley joins the main combe. The water emerges from a circular masonry-lined pipe, which can be followed for some distance, gradually diminishing in size. In days gone by it was reputedly frequented by local women who would drop pins into the water to aid fertility. A short distance upstream from the spring another strong flow of water emerges from an even smaller stone-lined conduit situated in the valley floor (ST 5611 5585) This water is presumably an overflow, derived from the artificial leat situated directly above, which feeds into the Line of Works. In 1830, a building was erected near here for washing the ore extracted from the local mines, and it is also said to be a favoured haunt of an ill-fated member of the de Gourney family, who haunts the wood dressed as a black-clad spirit. His is not the only malevolent presence in this vicinity, as a procession of black demonic beasts is also reputed to haunt nearby Spring Ground, close to the spot where suicides were once buried. The beasts apparently terrorise the locals witless, before vanishing into a nearby pool!

Ref: P. Quinn, The Holy Wells of Bath and Bristol Region, Logaston: Almeley, pp 175-176 (1999)

HARPTREE COMBE MINE 2 — Mine
ST 5606 5578   L 14m VR 3m Alt 125m

Also known as Twin Passage Mine and Mine 4002, this abandoned calamine working is located on the west bank of the combe, close to the valley floor. It consists of two interconnected passages, each 7m long, one of which is said to contain an intermittent sump, partially obscured beneath a floor of boulders. The 'sump' was not apparent in 2015.

Ref: A. D. Oldham, The Mines of Harptree Combe, SVCC Ntr 1(2) pp 3-4 (1963)

RIFT MINE — Mine
ST 5609 5577   L 25m VR 5m Alt 130m

Also known as Harptree Combe Mine 3 and Mine 4003, this impressive working is located on the east side of the combe, directly opposite Harptree Combe Mine 2. The first 8m has collapsed open to the surface, but the remainder is a roomy underground rift. Microscopic analysis of the smithsonite found inside this mine revealed cavities within the mineral encrusted with allophane and white aluminium.

Ref: A. D. Oldham, The Mines of Harptree Combe, SVCC Ntr 1(2) pp 3-4 (1963)

HARPTREE COMBE MINE 4 — Mine
ST 5609 5576   L 8m VR 3m Alt 128m

Also recorded as Mine 4004, this working is located on the east side of the combe a short distance up-valley from Harptree Combe Mine 3. A roomy opening, with a jammed boulder, yields a short mined passage. There is another tiny rift (L 2m VR 2m) situated in the same vein directly above.

Ref: A. D. Oldham, The Mines of Harptree Combe, SVCC Ntr 1(2) pp 3-4 (1963)

HARPTREE COMBE MINE 5 — Mine
ST 5607 5575   L 10m VR 5m Alt 125m

Also known as Mine 4005, this roomy, abandoned calamine working is located close to valley floor level a short distance up-valley from Harptree Combe Mine 4.

Ref: A. D. Oldham, The Mines of Harptree Combe, SVCC Ntr 1(2) pp 3-4 (1963)

HARPTREE COMBE MINE 6 — Mine
ST 5605 5578   L 13m VR 5m Alt 130m

Also known as Mine 4006, this unroofed calamine working is located on the steep bank immediately above Harptree Combe Mine 2.

Ref: A. D. Oldham, The Mines of Harptree Combe, SVCC Ntr 1(2) pp 3-4 (1963)

HARPTREE COMBE FIELD COLLAPSE — Collapse
ST 5595 5580   L 1m VR 1m Alt 142m

First noted in 2014, this small earthy collapse in a field above Harptree Combe may be related to mining in the combe below.

Ref: V. Simmonds, Personal communication

HARPTREE COMBE MINE 7 — Mine
ST 5608 5574   L 3m VR 2m Alt 130m

This narrow, mined rift is situated on the slopes directly above Harptree Combe Mine 5. There is another mined cavity in the castle rampart immediately above.

RICHMONT CASTLE RIFT — Mine
ST 5608 5569   L 10m VR 4m Alt 130m

Also logged as Harptree Combe Mine 8, this open mined rift lies on the edge of the rampart close to the footpath which descends from the castle down into the combe.

HARPTREE COMBE MINE 9 — Mine
ST 5601 5564   L 5m VR 2m Alt 130m

Another small abandoned calamine working.

HARPTREE COMBE SEEPAGE — Seepage
ST 5593 5543   Alt 135m

A tiny seepage next to the track in the valley floor.

HARPTREE COMBE MINE 10 — Mine
ST 5581 5539   L 5m VR 2m Alt 150m

A small pit yields a small mined rift. It is heavily choked with leaf debris but appears to quickly peter out in both directions. The first official record was made by WCC in 2011.

HARPTREE COMBE MINE 11 — Mine
ST 5578 5537   L 10m VR 4m Alt 150m

A roomy entrance, partly obscured beneath a large rhododendron, yields a steeply-descending, abandoned calamine working. It was first recorded by WCC in 2011.

BORDERLINES

RIDGE LANE SWALLET
ST 5497 5584
Dig
L 4m VR 4m Alt 221m

A short shaft excavated by local cavers during 1989-1990. No sign of a cave ever materialised although dowsing by Tony Blick of the CPC suggested a passage trending west. The shaft, and much of the depression, has now been backfilled. 300m to the north, in a field close to the hamlet of Ridge (formally Rudge) is a rectangular stone-walled enclosure which was an old cock fighting pit.

BORDERLINES DIG
ST 5470 5575
Dig
L 12m VR 12m Alt 237m

Dug by WCC in the early 1990s, no sign now remains of this depression which was excavated using a mechanical digger. No solid rock was ever found.

YOGHURT POT
ST 5444 5558
Cave
L 10m VR 3m Alt 249m

Located at the edge of the field boundary, adjacent to the main road, this 'gurt pot' was opened by WCC using a mechanical digger in 1991. Sadly the way on swiftly became too tight to follow. The muddy passage is now entered via a short section of horizontal concrete tubes, occasionally inhabited by foxes.

LAMB BOTTOM

Mining in this vicinity probably began in the mid-16th century and continued right through until the 19th century. The surrounding fields are pitted with old lead workings, many of which were backfilled with sawdust. During the 17th century, lead miners discovered Lamb Leer Cavern, a natural cavity containing a vast chamber, which for many years was regarded as the world's deepest explored cave.

LAMB LEER QUARRY RIFT
ST 5441 5519
Cave
L 12m VR 12m Alt 250m

Located at the east face of the quarry, this impenetrable vertical rift was dug briefly by UBSS in 1947. Tratman lowered a lantern down to deduce the true depth but failed to reach a definitive floor. The entrance now lies buried several metres below the present day quarry floor.

Ref: Old registry sheets, UBSS Hut Logs (1946-47) & N. Barrington, W. Stanton, The Complete Caves and a view of the hills, p 106 (1977)

GIBBETS BROW SHAFT
ST 5431 5510
Mine / Cave
L 130m VR 24m Alt 259m

Also known as the Shepton Secret Dig, this 19th century lead trial mineshaft gains a series of muddy phreatic tubes which lie directly above known passages in Lamb Leer Cavern. Scratched at by MNRC during 1959-60, the majority of the 19m deep shaft and the tubes beyond were excavated by SMCC (starting in 2003), and work continues sporadically to engineer a connection.

Ref: Mendip Underground, MCRA p 150 (2013)

LAMB LEER CAVERN
ST 5432 5505

Cave
L 585m VR 67m Alt 252m

Also recorded as Lamb Leer, Lamb Lair Cavern, Lamb Cave and Beaumont's Hole, this was once the world's deepest explored cave, a title that it held for a surprising 80 years. Discovered by lead miners, '*some years*' before 1676, a partly mined entrance shaft yields a low passage leading to chamber dominated by a large 'beehive' boss. Beyond Beehive Chamber, an artificially enlarged passage passes beneath a thick aragonite 'bridge' (the remains of a false floor largely destroyed by 17th century miners), then opens dramatically at a windlass platform, perched 20m above the floor of an enormous fault-aligned chamber. Great Chamber is over 30m high, about 20m in diameter, and two significant extensions lead off. The extensive St Valentine's Series is reached by a 10m vertical ascent, while a sloping boulder floor and 5m climb up a scaffold and ladder 'tower' gives access to Beaumont's Drive. This leads to the Cave of Falling Waters, containing a small stream and the well-known inscription 'T.W. 1894'. The initials were carved by Thomas Willcox, and the date added by a disapproving Herbert Balch, in an attempt to establish the rate of stalagmite deposition. WWA dye-traced the stream to Rickford Rising in 1977 (58 hours). The cave was first described by John Beaumont in Philosophical Collections No.2 (1681) and Beaumont's Drive, commemorates his name. He also claimed that his men sank a shaft 10 fathoms deep (20m), in the floor of Great Chamber. Beaumont lived at Upper Hay Street Farm in Ston Easton, and as the only practicing catholic in his parish, was forced to build his own chapel, which survives to this day, but is now used for other purposes. The original entrance shaft was 'lost' in the mid-1700s, but another was sunk in June 1880 by miners prospecting for iron ore, who also installed a winch at the head of the pitch into Great Chamber. The cave enjoyed a certain notoriety amongst the more adventurous ladies and gentlemen of the day. Closed by an entrance collapse in 1918, it was reopened in 1936, once again utilising Beaumont's original shaft. In 1937, WCC constructed a cableway across the Great Chamber to provide a spectacular traverse by cable car, but by WWII the system had fallen into disrepair and was removed. In 1938, Professor L S. Palmer carried out earth resistivity tests above the cave and the results suggested a second large void near Great Chamber. Further tests in 1958 placed the void 120m to the north-west of Great Chamber. MNRC maypoled into St Valentine's Series in 1958 and explored it in stages between 1961 and 1965. More resistivity tests were carried out by SMCC in 2005, but so far, 'Palmer's Chamber' has eluded discovery. It is thought that the earlier results may have been distorted by the presence of a local fault. The term

The Great Chamber of Lamb Leer Cavern.
Roy McPearce collection, courtesy of Wells and Mendip Museum

Leer is unusual and may have been introduced by German miners, who were pre-eminent in Europe at the time. However a more likely derivation is the Anglo-Saxon word 'lear', which means hollow or empty. In 1937, an interview with an 83 year old former miner named Andrew Lyons, revealed details of an underground lake, which has now been lost to us. During the 1880 re-discovery of the cave, Lyons described finding '*a gurt lake*', beyond a small passage apparently located close to the foot of the Great Chamber pitch. He made a raft out of a piece of wood, placed lighted candles on it and pushed it out across the water but he failed to discern any walls. Jack Duck, who interviewed him, said that the rest of Lyons' description of the cave was very accurate and he firmly believed that the old man had seen some sort of lake or large pool. A dig, Murrell's Dig, was started at the supposed spot, but this proved inconclusive. In 1946, Stanton recorded two mineshafts in the floor of the Great Chamber, and a tiny passage below the north wall which ended in a pool. Sadly the landowner closed the cave during the 1980s and the entrance is currently blocked.

Ref: Mendip Underground, MCRA pp 189-191 (2013)

LOST CAVE OF HARPTREE
Lost Cave

Purported to lie somewhere in the vicinity of Lamb Leer Cavern, local legend holds that miners entered another cave which consisted of a '*600 feet crawl to a decorated chamber*', a story related to local caver Vince Simmonds by villager Jack Lyons. However this story bears a strong resemblance to an earlier tale told by former miner Andrew Lyons (presumably a relative), to Jack Duck concerning the existence of a chamber containing '*a gurt lake*', allegedly located close to the foot of the Great Chamber pitch in Lamb Leer. As neither cavity has ever materialised, it may be that both tales are simply 'hand-me-down' echoes surrounding the loss of Lamb Leer Cavern during the mid-1700s.

GOODING'S 1
Mine

ST 5447 5505 approx.   L 20m VR 13m Alt 251m

Visited by ACG in 1970, this mineshaft was situated about '*200 yards*' from road and '*more or less due east from Lamb Leer entrance*'. It consisted of a 13m deep shaft cut in solid rock, which yielded a mined-out, muddy chamber containing two waterworn cross rifts. There are numerous backfilled mineshafts in this area and the precise site can no longer be identified.

Ref: ACG Logbook No.4 pp 53-54 (1970)

LAMB BOTTOM RIFT
Mine

ST 5445 5500   L 9m VR 4m Alt 245m

Also logged as Lamb Bottom Cave 2 and The Rift, this is the most obvious remnant of the considerable amount of mining that once took place in Lamb Bottom. The large unroofed rift was recorded by the BEC in 1961.

Ref: Late Entry to Caving Log, BEC Belfry Bull 157 pp 11-12 (1961)

LAMB BOTTOM ADIT
Mine
ST 5447 5495
L 9m VR 2m Alt 235m

Located directly opposite Lamb Bottom Rift, this conglomerate earth-roofed rift was visited by the BEC in 1961, who recorded it as Lamb Bottom Cave 1. They described the hole as *'about 30 ft long, 6-7 ft wide and 6 ft high'*. It was probably the remains of an old mining adit, although at least one large stalactite was discovered in the fill. It collapsed soon afterwards, before being partially reopened by BCSS in 1963. All that remains now is an earth filled cleft.

Ref: Late Entry to Caving Log, BEC Belfry Bull 157 pp 11-12 (1961)

HYWEL'S HOLE
Cave
ST 5446 5494
L 6m VR 5m Alt 241m

Also known as Hywel's Cave, this cave was apparently located 5m above valley floor level, on the left side of the bluff directly opposite both a notable pinnacle, and a mined gully (presumably Lamb Bottom Rift). It was first entered by Hywel Murrell in 1934, but was covered over with a stone slab and lost shortly after. Murrell relocated it in 1963, and it was found again by WCC in 1966. The cave consists of a drop into a steeply-sloping, choked waterworn fissure. No obvious opening exists now, although a small square-shaped depression, situated directly above the remains of Lamb Bottom Adit, may be the best candidate. Hywel also scratched at the marshy hollow in the valley floor nearby.

Ref: N. Barrington, W. Stanton, The Complete Caves and a view of the hills, p 101 (1977)

LAMB BOTTOM FISSURE
Cave
ST 5456 5489
L 2m VR 1m Alt 230m

Recorded on old registry sheets as both Garrow Fissure and Impenetrable Fissure, this low bedding is located beneath the last limestone bluff on right-hand side of Lamb Bottom when travelling down-valley. Burr also mentions a fissure, filled with iron ore and measuring 2m across, somewhere in this vicinity.

Ref: P. Burr, Mines and Minerals of the Mendip Hills, Volume 1, MCRA p 249 (2015)

GARROWPIPE FISSURES
Caves
ST 5487 5484
L 5m VR 2m Alt 215m

The Dolomitic Conglomerate cliff above the Garrowpipe Springs is pierced with several small fissures, some of which interconnect. They may have been mined. Ray Mansfield recorded these as Upper and Lower Garrow Caves.

Ref: R. Mansfield, Caving Diary and General Notes, Vol 2 p 3 (Apr-Dec 1963)

GARROWPIPE SPRING
Spring
ST 5485 5487
L 2m VR 1m Alt 200m

This lively spring rises from beneath a pile of huge Dolomitic Conglomerate boulders near the head of Garrow Bottom. A dig by the BEC in 1972, at a point midway between the spring and the cliff face above, yielded a brief section of the subterranean stream. There are no proven feeders, but water sinking in and around 'The Belt' appears to be the most likely source. In November 1887, while searching for the source on land to the west of Frances Plantation, a local labourer called William Currell, put his pick into a pewter vessel full of nearly 1500 Roman coins with five ingots of silver and a ring. The jar lay six inches below the surface in swampy ground. The coins, which were all struck around AD 375, became known as the 'Harptree Hoard', and many were given to the British Museum, where they can still be seen today. The remainder were kept in the jar in East Harptree church, from which, unfortunately, they were stolen in the 1970s.

Ref: N. Barrington, W. Stanton, The Complete Caves and a view of the hills, p 85 (1977)

GARROWPIPE SPRING 2
Spring
ST 5492 5486 Alt 200m
This small spring lies roughly 65m east of the main Garrowpipe Spring, at the same level. In wet weather a substantial seepage appears in the valley floor nearby.

SMITHAM HILL & EAST HARPTREE WOODS

SMITHAM HILL COLLAPSE
Sink
ST 5547 5360 L 2m VR 2m Alt 292m
Two collapses on the north side of a complex depression. The lower collapse takes a wet weather stream while the upper collapse is gradually being filled with rubble.

SILAGE TRACTOR COLLAPSE
Collapse
ST 5536 5396 L 3m VR 3m Alt 285m
Situated in a cluster of shallow depressions, this small collapse appeared in 1985 courtesy of a tractor hauling a silage trailer. Probably an old ochre pit, it consisted of a small, waterlogged dome chamber floored with a debris cone of chert blocks and ochreous mud. It was quickly filled in.

Ref: W. I. Stanton, Logbook 17 pp 48-49 (1985)

SMITHAM HILL RIFT CAVE
Cave
ST 5572 5402 L 5m VR 4m Alt 272m
A tight vertical rift, situated at the eastern end of a small depression in the woods, yielded a loose, muddy chamber formed in Harptree Beds. It was dug by the BEC in 1991-92 and was notable for the live flies that were seen to emerge from inside the rocks as they were broken up! The cave is now completely backfilled.

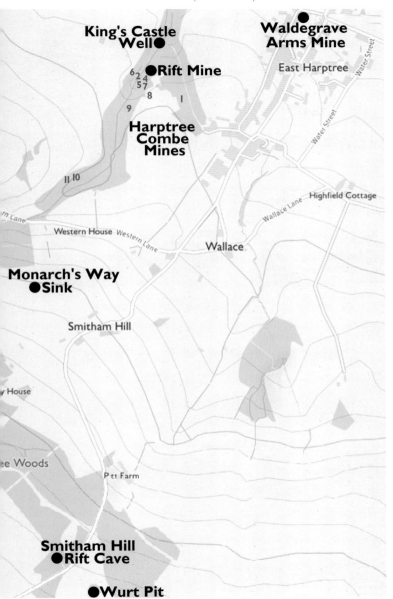

HAWTHORN HOLE
Collapse
ST 5582 5373
L 6m VR 6m Alt 283m

This natural clay collapse occurred in a hedgerow shortly after the 1968 floods. Situated adjacent to a deep depression it was swiftly filled in and is now barely recognisable.

Ref: N. Barrington, W. Stanton, The Complete Caves and a view of the hills, p 94 (1977)

WURT PIT
Doline / Cave
ST 5587 5391
L 5m VR 2m Alt 274m

Also known as Wort Pit (presumably after the whortleberries which grow around the sides), this huge shakehole is possibly the largest such feature on Mendip. An important SSSI, it features good exposures of silicified clays, shale and Lias limestone, known locally as Harptree Beds. These silicified layers are a surviving remnant of the soft lower Jurassic strata which is still extant in the far east of Mendip, but which has been almost entirely removed further west. The silicification process (saturation of organic matter with silica), has rendered the Jurassic strata here significantly more resistant to weathering, and has been attributed to hydrothermal activity caused by hot, silica-rich fluids rising up the Biddle Fault. This probably took place as part of the main phase of mineralisation which emplaced the principal Mendip ore-fields during the Jurassic. The Harptree Beds show varying degrees of silica-enrichment, and often contain traces of important minerals, including limonite, ochre, lead and zinc. Locally, the beds overlie Dolomitic Conglomerate, possibly sandwiching a thin layer of Keuper Marl, and the depression is probably the result of solution of the conglomerate at depth, which has led to the collapse of the layers above. There is also a short bedding cave, located close to the northern rim. Often home to animals, this has largely been filled in. In 1968, contamination of springs in this vicinity caused significant pollution in Chew Valley Lake.

Ref: N. Barrington, W. Stanton, The Complete Caves and a view of the hills, p 181 (1977)

PITT FARM SPRING
Spring
ST 5597 5419
Alt 232m

The small spring-fed pond situated alongside the track leading to Nett Wood Farm, was probably used for washing the yellow ochre mined from the surrounding fields in the 19th century. The lively outflow from the pond vanishes behind Pitt Farm before reappearing a short distance further downhill. It is probably piped beneath the garden. Several springs rise in the field to the north-east of the farm complex, which may explain why it was recently rebadged as Springfield Farm. It was here that a horse called Buzz fell into a 2m deep sinkhole that suddenly appeared in 2017. Happily he was rescued relatively unscathed.

THE BELT

HARPTREE WOODS SINK
Sink
ST 5505 5436 L 1m VR 1m Alt 258m

The current western edge of East Harptree Woods was once a separate entity known as 'The Belt'. A number of swallets have developed in this vicinity and are probably responsible for much of the water resurging from the Garrowpipe Springs, although the Double or Canyon Springs remain other possibilities. Located below a woodland path and 20m south of an obvious footbridge, this small double depression has a number of drainage ditches running into it and at times it has been seen to take a substantial amount of water. The sink is developed within the Harptree Beds and there are exposures of this rock type evident in the depression sides. It was dug by the BEC in 2012, who reached a tiny chamber, but recent slumping prevents access. The Forestry Commission has directed drainage channels into a number of neighbouring depressions.

Ref: http://ramblingon.mendipgeoarch.net

FRANCES PLANTATION SWALLET
Sink
ST 5498 5441 L 3m VR 3m Alt 244m

A stream from the adjacent field sinks in a small depression located below two Ash trees at the very edge of the wood. It was dug briefly by WCC about 1965 who sank a 3m deep shaft through Red Marl. The shaft has since refilled, although the moss-covered spoil heap is still identifiable. The name is unfortunate because the sink is actually located in 'The Belt', a line of ash trees which were originally distinct from Frances Plantation.

Ref: N. Barrington, W. Stanton, The Complete Caves and a view of the hills, p 81 (1977)

BELT SINK SOUTH
Sink
ST 5490 5453 Alt 246m

A wet-weather stream sinks in a rubble-filled depression, located in the open field 60m west of the wood.

BELT SINK NORTH
Sink
ST 5490 5457 Alt 246m

Located 40m north of Belt Sink South, a wet-weather stream sinks in a depression shaped like a small blind valley.

POOL SINK
Sink
ST 5503 5451 Alt 250m

A spring, situated at the head of an elongated woodland depression, usually sinks in a small murky pool. In wet weather it overflows and runs along the surface to Garrow Bottom. The Forestry Commission has again directed drainage channels into several neighbouring depressions.

FRANCES PLANTATION

Frances Plantation is named after the famous Victorian hostess, Frances, Countess of Waldegrave. A charismatic character, she rescued Sir Horace Walpole's derelict Strawberry Hill residence, and once spent six months living inside Queen's Bench prison alongside her husband, the 7th Earl, who had been sentenced for assault.

CHOCOLATE WALL SINK   Sink
ST 5510 5466   L 1m VR 1m Alt 246m

A small stream trickles down an attractive banded wall of Red Marl to sink in the side of a large depression. This is almost certainly the site first noted by two London cavers in 1949, and may be the site of **Chimney Pot**, which was excavated for 4m by J. Hanwell and E. J. Maunders in 1954. Smitham Sink, or an unrecognisable backfilled location just north of the chimney have also been suggested as possible locations. The water probably re-emerges at the Double and Canyon Springs above Garrow Bottom.

Ref: P.A. E. Stewart, Some hitherto unrecorded expeditions and discoveries on Mendip, 1947, 1948, 1949, 1950, WNHAS & MNRC Report 63/64 p 14 (1951/1952)

SMITHAM SINK   Sink
ST 5519 5471   L 4m VR 4m Alt 244m

This small sink, which is situated in a relatively open glade close to a wall, was first dug by WCC in 1976 and has been probed by local cavers since. The rising for the water is unknown, but the Double and Canyon Springs above Garrow Bottom are strong possibilities.

Ref: N. Barrington, W. Stanton, The Complete Caves and a view of the hills, p 146 (1977)

CHIMNEY SINK   Sink
ST 5538 5480   Alt 230m

Located 180m due north of Smitham chimney and clearly marked as active on old maps, this was the main sink point for the pond outfall before it was diverted to its modern position adjacent to the big slag heap (**Slag Heap Sink**, ST 5550 5469). It probably still takes water in very wet weather, along with a very large depression (**Flat Bottomed Sink**, ST 5544 5481), which also appears on old maps. The water for all of these sinks probably feeds the Double and Canyon Springs above Garrow Bottom. The chimney is the sole Mendip survivor of several erected by the Cornishmen in the 19th century. Long threatened with demolition, it was saved in 1973 by the Mendip Society, acting in collaboration with Somerset County Council. All the other buildings were destroyed, although the ruins of five flues, similar to those still visible at Charterhouse and St Cuthbert's can be found in the undergrowth to the north-east.

CANYON SPRING
Spring
Alt 201m
ST 5543 5504
Several springs unite at the head of a small canyon. There are no proven feeders but the water sinking at Smitham Sink and Chocolate Wall Sink appear a likely source (see also Double Spring below).

DOUBLE SPRING
Spring
Alt 185m
ST 5535 5512
There are no proven feeders to this small double spring, although Smitham Sink and Chocolate Wall Sink may be possible sources.

GARROW BOTTOM SPRING
Spring
Alt 185m
ST 5525 5509
A small spring, situated just below the rusting remains of an old abandoned chassis (2015). There is another small spring nearby, at ST 5520 5507.

MONARCH'S WAY SINK
Sink
Alt 195m
ST 5568 5498
An intermittent trickle sinks in a large depression, where some recent tipping has occurred. It possibly reappears at a small spring located 140m down the hill (ST 5574 5508). The sink lies next to the Monarch's Way, a long distance footpath based on the route followed by King Charles II following his defeat by Cromwell's New Model Army at the Battle of Worcester in 1651. It was the last battle of the English Civil War, a series of bloody conflicts fought between Parliamentary forces (Roundheads), and Royalists (Cavaliers), from 1642 to 1651.

EAST HARPTREE & LITTON

HARPTREE COURT TUNNEL & GROTTO
Artificial Grotto
Alt 90m
ST 5684 5613
An artificial tunnel and grotto in the private grounds of Harptree Court. The house was built in 1797 but is perhaps most famous for hosting Series 3 and 4 of the Great British Bake Off.

HARPTREE FARM MINE
Mine
Alt 105m
ST 5669 5603
In 1904, workmen digging the foundations for new stables at Harptree Farm, uncovered a small lead working *'through which a man could walk under Rectory Lane towards the church'*. Unusually, it was stone-lined, presumably intended to help support the buildings above. The village may be honeycombed with similar small workings and records suggest that mining took place as early as 1728. Most however, such as those uncovered at Zion (Sion) Place, near the south-west corner of Church Lane, are believed to date to c.1800. They were probably sealed in the late 19th century.

Ref: Struggles of the poor in Mendip's lead mining villages, Weston Mercury (10th Dec 2007)

WALDEGRAVE ARMS MINE   Mine
ST 5659 5595   L 40m VR 5m Alt 110m
Revealed by a collapse in the pub car park in 1980, this small calamine and ochre working yielded three low tunnels containing old coins, metal barrel hoops and the bones of a goose. There were no less than seven choked shafts overhead, including the point of entry, a 5m shaft which was swiftly capped. A 40m deep well in the car park has also been backfilled.

Ref: W. I. Stanton, Logbook 15 pp 22-23 (1980)

EASTWOOD MANOR MINE 1   Mine
ST 5759 5512   L 10m VR 3m Alt 125m
Located in the south face of a small quarry which provided stone to build the modern Eastwood Manor, this short mine is entered by 2m wide slot.

Ref: V. Simmonds, Eastwood Manor Mines, BEC Belfry Bull 504 pp 23-24 (1999)

EASTWOOD MANOR MINE 2   Mine
ST 5757 5514   L 90m VR 5m Alt 125m
A low entrance in the west face of the quarry drops into roomy passage which runs parallel to the quarry face. There are several blocked passages and it is inhabited by badgers. Originally known as Capon Mine, it was a small barytes venture operated by the Kingsway Syndicate, which employed 10-15 men between 1906 and 1907. Red ochre was also excavated in this vicinity.

Ref: V. Simmonds, Eastwood Manor Mines, BEC Belfry Bull 504 pp 23-24 (1999)

ARM COVER LEAT   Tunnel
ST 5774 5514   L 50m VR 5m Alt 112m
Protected by a barrier and covered with a metal grill, this brick-lined tunnel is located in bushes opposite the pond at Lodge Cottages. Probably built to control the flow of water entering the millpond, it has been followed to a chimney and sluice.

Ref: W. I. Stanton, Logbook 6 p 29 (1949)

SHERBORNE SPRING   Rising
ST 5858 5504   Alt 92m
Also recorded as Sherbourne Spring, this strong clear spring, which rises from the Red Marl through fissures and gravel, feeds into the Bristol reservoirs at Barrow Gurney via the 'Line of Works'. Capped by Bristol Waterworks in 1882, the spring probably drains much of Greendown, West End and Smitham Hill, but the only proved feeders to date are Vee Swallet and All Eights Mineshaft. The name probably derives from 'scir burne', an Old English term meaning clear or radiant stream. In 1980 a small collapse (**Sherborne Spring Collapse**) occurred on the opposite side of the road at ST 5858 5493. Formed in Red Marl, it was 2m deep and 2m in diameter and was quickly filled in.

Ref: N. Barrington, W. Stanton, The Complete Caves and a view of the hills, p 143 (1977) & W. I. Stanton, Logbook 15 p 42 (1980)

EAKER HILL & GREENDOWN

Given the substantial number of shafts and spoil heaps which still litter the ground to this day, it is perhaps surprising that the well-preserved mining area to the north of Eaker Hill remains comparatively less well-known than other mining areas of Mendip. The thickest concentration of shafts lies in the vicinity of Eaker Hill Farm and Buddle's Wood, and there is clear evidence that these were worked over centuries, culminating in several deep shafts sunk by Cornishmen in the mid-19th century. Early 17th century mining records held by the Waldegrave Estate Office list regular returns from Chewton, Predy, Bousland, Hartrey and West. Chewton, Predy and Hartrey are Chewton, Priddy and Harptree mineries respectively, while West is synonymous with Charterhouse. The name Bousland however has not survived in the landscape and as such has never been positively identified. However, this author believes that Bous may be a corruption of Bois, the French word for wood, and as old maps imply that the other possible contenders were not wooded at that time, the most likely candidate is the area now occupied by Bendall's Grove and Buddle's Wood. Ore extracted from this area would have been transported down the hill to Tor Hole for processing. Eaker Hill is also the suggested location for Ager Hill, where a ghostly apparition was witnessed on 10th October 1835. Observers reported seeing '*an immense body of troops, mounted and fully accoutered ... with drawn swords at the carry*'. The apparition was viewed for '*upwards of an hour, and the cottagers around the foot of the hill, were, for a considerable time in a state of consternation*'. This extraordinary event was later attributed to refraction, which somehow managed to magnify the Bath Troop of Yeomanry Cavalry who had assembled at Twerton, some 15 miles away. However, this author contends that some of the local cider may have played a part, perhaps coupled with an old folk memory of an actual civil war skirmish said to have taken place in these parts in 1643.

EAKER HILL & WEST END

ALL EIGHTS MINESHAFT   Mine
ST 5596 5291   L 91m VR 46m Alt 282m

Also known as West End Mineshaft, Double Eights Mineshaft and Shaft 4089, this interesting, if somewhat overgrown mine, comprises two parallel, interconnecting vertical shafts, one round, one square, which reach water level at -37m (total depth is said to be -46m). There is only one side passage, an unstable stope to one side, and the water, which can vanish completely during drought conditions, is claimed (dubiously) to have been dye-tested to Sherborne Spring, almost 3.5km away. The round shaft (now blocked) was probably the oldest and may be the site opened c.1822, as part of the conditions for allowing lead to be worked near Wigmore Farm. It was proposed that a fire-engine should be erected to help drain water from a mine known as 'Engine Pit'. The splendid square shaft was probably sunk either by the Mendip Hills Mining Company c.1853, or alternatively by W. Thomas c.1867. Old maps show a winding house present on the site, and wire rope and engine footings were still clearly visible in 1939. There are extensive surface workings visible in the wood immediately to the west, while the numerous spoil heaps in the fields to the east pay testament to the literally dozens of similar workings which must have once littered this hillside. Only three other shafts remain open, and these were visited by the BEC in 1993 (see below). Of the others, three (5m, 7m, and 12m), were descended by Gerard Platten, but these cannot be positively identified.

Ref: J.W. Gough, The Mines of Mendip, David & Charles p 176 & p 247 (1967)

ALL EIGHTS FIELD SHAFT 1 — Mine
ST 5608 5288   L 7m VR 7m Alt 282m

Logged on early registry sheets as Shaft 4090, this heavily overgrown fenced shaft is 7m deep to a blockage with only one tiny side passage. The shaft was first described by H. Murrell in 1936, and was descended by BEC in 1993, who noted smoke marks on walls.

Ref: A. Jarratt, Logbook 5 p 111 (1993)

ALL EIGHTS FIELD SHAFT 2 — Mine
ST 5607 5280   L 18m VR 8m Alt 282m

Logged on early registry sheets as Shaft 4091, this overgrown fenced shaft is 8m deep to a blockage in the floor, and a 5m stope on either side. It was descended by BEC in 1993, who noted pick and chisel marks in the clay, and good slickensides on the smoke-marked walls. This is almost certainly the shaft visited by H. Murrell in 1936, who described a shaft '*40 feet deep*' which crossed a partially walled-up natural rift.

Ref: H. Murrell, MS Diary 3,52 & A. Jarratt, Logbook 5 p 111 (1993)

WEST END OLD SHAFT 1 — Mine
ST 5637 5293   L 23m VR 23m Alt 271m

In 1956, WCC explored a '*70 feet deep*' shaft uncovered by a bulldozer. It led to shallow water, with an unexplored side passage. In 1949, Stanton recorded a shaft '*a few hundred yards east of All Eights Mineshaft*'. A tight entrance yielded a 7m shaft to a narrow thrust-plane, which also was descended to water, reaching a total depth of 20m. Here too, a narrow side joint remained unexplored. It is not clear if this is the same site.

Ref: W. I. Stanton, Logbook 6 p 104 (1949) & WCC Logbook 1955-56 p 104 (1956)

F & B HOLE
Shaft
ST 5659 5290
L 8m VR 8m Alt 265m

This grilled shaft, located on the south-east corner of the derelict Eaker Hill Farm, has been dug intermittently by WCC since c.2014. It is considered to be natural, but given its location was probably modified by miners. F & B stands for Faggot and Beans, in tribute to the late Tony Jarratt, who often ate them in The Hunters' after digging!

EAKER HILL FARM MINESHAFT
Mine
ST 5670 5283
L 30m VR 30m Alt 262m

Marked as a Trial Shaft on OS maps, and recorded as Mineshaft 4124 by Mansfield, this shaft, enclosed within a circular steel fence, is reputed to have supplied water to the farm. H. Murrell first noted it in 1936, but was unable to descend as it was covered by stone slabs deemed too heavy to move. It was eventually bottomed by BEC in 1963, who reached a boulder floor with no way on at 30m depth, and was visited again by the BEC in 1993, who found it choked with sheep carcasses at -19m. There are numerous capped or backfilled shafts in this intensively mined area, one of which **Mineshaft 4092** (ST 5663 5279), was descended by H. Murrell for 7m to a choke. Others appear on old maps of the area, including **West End Old Shafts 2** (ST 5675 5278). Stalagmite found in a dump in the north-east corner of the field (ST 5689 5278), suggests that the miners encountered at least one cavern somewhere in this vicinity.

Ref: W. I. Stanton, Logbook 6 p 104 (1949), R. Mansfield, Caving Diary and General Notes, Vol 2 p 14 (Apr-Dec 1963) & A. Jarratt, Logbook 5 p 111 (1993) & N. Richards, Personal communication

BENDALL'S GROVE & BUDDLE'S WOOD

To the north of Eaker Hill, there are two small wooded areas (Bendall's Grove and Buddle's Wood) in the private Waldegrave Estate, both of which have been extensively mined. Choked shafts of up to 20m in diameter are commonplace on the higher slopes, often surrounded by distinctive and well-preserved collars of waste material. A handful of larger shafts may have been sunk by MHMCo c.1853, but the majority probably date from much earlier. There are no open shafts, or even historical records of the mines in Bendall's Grove, but over the years several shafts in Buddle's Wood have been descended, including a few visited by Murrell during the 1930s. A number of those were subsequently filled in, only to be replaced by sixteen others, which burst open when floodwater rushed down through the wood during the 1968 flood. Many of the shafts are located in a relatively narrow band on either side of what was, until recently, a firebreak path. However, the trees to the east of the path have since been felled, and that area is now so overgrown that attempts

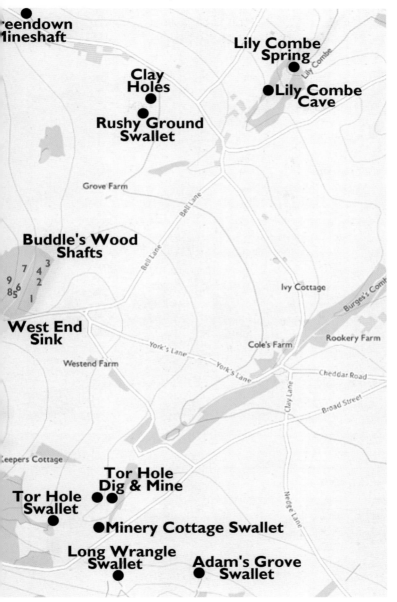

A typical mineshaft in Bendall's Grove. Photo by Rob Taviner

to identify some of the shafts recorded in the earliest descriptions remain somewhat open to interpretation. All surviving shafts were surveyed relatively recently by the UBSS, acting on behalf of the Waldegrave Estate (Buddle's Wood Survey Project).

WEST END SINK

Sink
ST 5725 5275
Alt 220m

This tiny trackside stream, which disappears into a road drain, dries up in summer but in wet conditions can flow across the road before finally vanishing down Buddle's Wood Shaft 1. Part of the water is derived from **West End Spring** (ST 5712 5266 Alt 235m), and part from **Soft Well** (ST 5699 5271 Alt 237m), which in times gone by presumably provided water both to the mines, and to nearby Eaker Hill Farm (now a forlorn ruin). West End Spring is interesting as it rises in the floor of an artificial brick-lined chamber. The water is then piped into a lower brick-lined chamber, where it normally sinks in dry weather.

BUDDLE'S WOOD SHAFT 1

Mine
ST 5735 5284
Alt 213m

No details are known about the three collapses blocked with tree trunks contained within this large fenced-off area. Clearly marked on old maps, this, the most southerly of three large shafts located along the eastern edge of the wood, swallows the wet weather overflow from West End Sink, and is significantly larger than the shafts located further west. It is probably one of the Cornish shafts sunk by MHMCo c.1853.

Ref: A.A. D. Moody, Personal communication

BUDDLE'S WOOD SHAFT 2

Mine
ST 5738 5289
L 23m VR 23m Alt 212m

The middle one of the three Cornish shafts located along the field boundary, this mine was visited by Hywel Murrell in 1936, who descended a '*77 ft deep*' shaft to a natural rift. Also logged as Buddles Pit and Mineshaft 4117, a heading led to further passages including a second shaft that was not explored due to the threat of collapse. It was revisited by SMCC in 1955, who found the shaft to be blocked at a depth of 17m, and it was subsequently completely filled in. There is a large mound in the field nearby which may be all that remains of another shaft, or possibly a spoil heap

Ref: H. Murrell, MS Diary 3 p 44, R. Mansfield, Caving Diary and General Notes, Vol 2 p 9 (Apr-Dec 1963), R. Biddle, Extracts from the S.M.C.C. Hut Logs, Part One, Volumes 1 and 2: April 1955 - May 1958 (1968) & A.A. D. Moody, Personal communication

BUDDLE'S WOOD SHAFT 3

Mine
ST 5741 5295
L 20m VR 20m Alt 208m

This large open shaft, surrounded by a circular fence, is the most northerly of the three Cornish mines clearly marked on old maps. Descended to a choke at -10m in 1982, it was noted by SMCC in 1963 and may well be Mineshaft 4119, a shaft first visited by Murrell in July 1936, who descended for '*65 ft*' without reaching bottom. It was apparently visited again by a joint EDCG/MNRC party in 1963, but was found to be partially blocked with boulders at '*-50 ft*'.

Ref: H. Murrell, MS Diary 3 p 52, R. Mansfield, Caving Diary and General Notes, Vol 2 p 10 (Apr-Dec 1963), M. Ellis, Extracts from the S.M.C.C. Hut Logs, Volume Four: July 1960 - December 1964 (1982), SMCC Jnl 9 (3) p 162 (1982) & A.A. D. Moody, Personal communication

BUDDLE'S WOOD SHAFT 4 — Mine
ST 5737 5293   L 10m VR 10m Alt 212m

Located in the middle of the scrubby ground to the east of the path, this shaft has a square fence and is grilled. In 1982 it was descended to a choke at -10m, and is another possible contender for Mineshaft 4119, the deep shaft first visited by Murrell in July 1936 (see Shaft 3 above).

Ref: H. Murrell, MS Diary 3 p 52, R. Mansfield, Caving Diary and General Notes, Vol 2 p 10 (Apr-Dec 1963) & A.A. D. Moody, Personal communication

BUDDLE'S WOOD SHAFT 5 — Mine
ST 5728 5283   Alt 221m

The most obvious of the shafts listed by the UBSS Buddle's Wood Survey Project, this small triangular enclosure, situated alongside the firebreak track, protects an unexplored shaft currently blocked by a large boulder or tree stump. This is almost certainly the site visited by a joint EDCG/MNRC party in 1963 and recorded as Mineshaft 4115.

Ref: R. Mansfield, Caving Diary and General Notes, Vol 2 p 9 (Apr-Dec 1963) & A. A. D. Moody, Personal communication

BUDDLE'S WOOD SHAFT 6 — Mine
ST 5729 5284   L 26m VR 12m Alt 221m

Located alongside the track, 20m north of Shaft 5, this open shaft is grilled and fenced with a square enclosure and is almost certainly Mineshaft 4118, which was first visited by Murrell in 1936. It currently consists of a 10m deep shaft, landing on a pile of deads and rubbish, with a visible continuation below. An ascending passage, entered by a traverse part way down the shaft, leads to a ginged shaft blocked by a capstone, which corresponds with a blocked shaft located a few metres to the south. The side passage contains impressive rope groove marks. Murrall described it as '*a 50ft shaft to a natural rift*' and it was revisited by SMCC in 1963, who descended the shaft for 40ft (the bottom was visible 10ft below), and explored a natural passage running upwards for 25ft. The shaft they described was mainly natural, but contained cavities filled with deads.

Ref: H. Murrell, MS Diary 3 p 52, R. Mansfield, Caving Diary and General Notes, Vol 2 p 10 (Apr-Dec 1963), M. Ellis, Extracts from the S.M.C.C. Hut Logs, Volume Four: July 1960 - December 1964 (1963), SMCC Jnl 9 (3) p 162 (1982) & A.A. D. Moody, Personal communication

BUDDLE'S WOOD SHAFT 7
ST 5733 5293
Mine
L 185m VR 18m Alt 220m

Located in woodland a short distance above the track, the majority of this mine was rediscovered by WCC in 1982. An 8m deep shaft, which is enclosed by a small square fence and covered with a wooden crate, yields a bouldery descent to a small chamber walled with deads. Beyond, a complicated network of small passages can be explored, one containing a narrow band of galena, and another sporting footholds apparently designed for children. It may correspond to Mineshaft 4116, a mine visited by a joint EDCG/MNRC party in 1963, who described a '*35 ft*' deep shaft, located '*160 yards up the fire break*' on the west side, and '*ten yards from track*'. This yielded a tight natural rift '*60 feet long*', with a loose boulder mass in the floor of the shaft which indicated that it was probably once deeper. However the location could just as easily refer to one of the many other choked shafts in the wood, such as the small unfenced hole located a few metres away (**Buddle's Wood Shaft 7B**).

Ref: R. Mansfield, Caving Diary and General Notes, Vol 2 p 9 (Apr-Dec 1963) & A. A. D. Moody, Personal communication

BUDDLE'S WOOD SHAFT 8
ST 5727 5283
Mine
L 4m VR 4m Alt 225m

Located in woodland immediately above Shaft 5, this site actually consists of three separate grilled shafts, recorded by UBSS as Shafts 8A, 8B and 8C. One was still open in 1982, when a tight 4m shaft was descended to a choke, with daylight visible entering from one of the other shafts. 8A and 8B are located in one fenced enclosure, and 8C in another, and all are now blocked.

Ref: A. A. D. Moody, Personal communication

BUDDLE'S WOOD SHAFT 8.5
ST 5726 5287
Mine
L 15m VR 5m Alt 227m

Dug open by UBSS during the course of their survey, this grilled 5m deep ginged shaft lies in woodland on the west side of the track and is enclosed by a square fence. Below, a short section of steeply-ascending passage remains accessible.

Ref: A. A. D. Moody, Personal communication

BUDDLE'S WOOD SHAFT 9
ST 5727 5290
Mine
L 600m VR 30m Alt 235m

Almost certainly the site recorded as Mineshaft 4121, this mine was first visited by Murrell in 1936, who described a '*40ft deep shaft*' to a natural rift. It was visited again by a joint EDCG/MNRC party in 1963, who explored a natural rift '*50 ft*' long containing many recent bones. There was a small nest of cave pearls in an offshoot at the far end of the rift, and '*12 ft*' above the floor. Most of the cave however was explored by WCC in 1981-82, who passed an exceedingly tight squeeze at the bottom of the pitch. This yielded multiple levels of partly natural passages containing excellent examples of cave pearls, rope grooves, and flowstone coated shot-holes.

Ref: H. Murrell, MS Diary 3 p 52, R. Mansfield, Caving Diary and General Notes, Vol 2 p 10 (Apr-Dec 1963) & A.A.D. Moody, Personal communication

BUDDLE'S WOOD OUTER SHAFT Mine
ST 5714 5288 Alt 247m

This very loose shaft, which is fenced and covered by a fallen tree in the firebreak separating Buddle's Wood from Bendall's Grove, has been descended for only a few metres, although rocks have been heard to drop for a considerable distance, hitting the bottom with a sizeable boom! There are several shafts in this vicinity, and the fact that the area is clearly marked on old maps, suggests that the shafts are Cornish in origin. This is probably the site Mansfield recorded as Mineshaft 4113. There is also an interesting pile of rocks at ST 5715 5293, which is presumably covering another old shaft.

Ref: A.A.D. Moody, Personal communication & R. Mansfield, Caving Diary and General Notes, Vol 2 p 11 (Apr-Dec 1963)

MINESHAFT 4120 Mine
ST 5733 5292 Alt 220m

This unexplored shaft was visited by Murrell in 1936, who found the top covered by an immovable slab. By 1963 it had been completely covered in earth. This may correspond to the obvious depression located immediately below, and to the east of the track.

Ref: H. Murrell, MS Diary 3 p 53 & R. Mansfield, Caving Diary and General Notes, Vol 2 p 10 (Apr-Dec 1963)

MINESHAFT 4122 Mine
ST 5734 5306 L 15m VR 13m Alt 216m

Visited by Murrell in 1938, this was described as a 40ft deep shaft to a well decorated grotto. The shaft had been completely filled in by 1963. The outlines of other infilled shafts are visible in the same field, along with several distinctive humps in the ground, which indicate the locations of vanished buildings marked on old maps.

Ref: The lost winter record of H. Murrell & R. Mansfield, Caving Diary and General Notes, Vol 2 p 11 (Apr-Dec 1963)

MINESHAFT 4123 Mine
ST 5740 5310 Alt 212m

An unexplored mineshaft, noted by Murrell in 1936. It had been filled in by 1963.

Ref: H. Murrell, MS Diary 3 p 52 & R. Mansfield, Caving Diary and General Notes, Vol 2 p 11 (Apr-Dec 1963)

GREENDOWN

GREENDOWN MINESHAFT
Mine
ST 5738 5384　　　　　　　　　　　　　　　L 13m VR 13m Alt 198m

A large square-walled shaft, which is currently choked with debris after 13m. A visit in 1975 found it floored with house debris with a blocked rift leading off. It may have been sunk by MHMCo c.1853, along with several other blocked shafts which are clearly visible both in this field, and in the neighbouring field to the north-west. In the middle of this 'mine field' there is a mysterious feature marked as a 'grotto' which appeared on maps up until the 1960s (ST 5738 5388). Apparently grubbed out, its purpose remains unclear although it was presumably an artificial garden folly.

Ref: R. Mansfield, Caving Diary and General Notes, Vol 7 July 26th (1975)

RADFORD FARM SINK
Sink
ST 5764 5373　　　　　　　　　　　　　　　L 1m VR 1m Alt 197m

A tiny sculpted rift beneath the lawn which takes storm water from the farmhouse. It was deemed too tight to pursue when examined briefly by WCC and ATLAS in 2001.

CLAY HOLES
Cave
ST 5779 5359　　　　　　　　　　　　　　　L 12m VR 8m Alt 199m

Usually flooded to within 3m of the surface, this large, steeply-descending passage was dug open by WCC in 1992-1996. It peters out in small mud-filled tubes.

Ref: R. G. Witcombe, Around the Wessex Digs, WCC Jnl 22 (241) p 94 (1994) & Jnl 23 (251) p 139 (1996)

RUSHY GROUND SWALLET
Cave
ST 5774 5352　　　　　　　　　　　　　　　L 59m VR 15m Alt 195m

This curious phreatic cave consists of a low undulating passage with a few minor offshoots. It contains some bone deposits and a few good formations. The cave was opened by WCC and landowner Jim Young in 1992.

Ref: Mendip Underground, MCRA pp 283-284 (2013)

LILY COMBE CAVE
Cave
ST 5817 5356　　　　　　　　　　　　　　　L 53m VR 24m Alt 168m

This low, steeply-sloping bedding plane was excavated by Rentahole CC in 1972-74 who even installed mains lighting! Sadly it was covered by demolition rubble c.1990 when the valley was given over to landfill.

Ref: N. Barrington, W. Stanton, The Complete Caves and a view of the hills, p 107 (1977)

LILY COMBE SPRING
Spring
ST 5825 5364　　　　　　　　　　　　　　　L 4m VR 4m Alt 155m

This intermittent spring feeds a large circular pond at the foot of Lily Combe. In 1973, Rentahole CC dug a 4m deep shaft in dry conditions - before the valley was infilled and the pond created. The flow was said to affect the water levels in Lily Combe Cave.

Ref: N. Barrington, W. Stanton, The Complete Caves and a view of the hills, p 107 (1977)

CHEWTON MENDIP

CHEWHILL HEAD RISING
ST 6003 5318

Rising
Alt 143m

Also known as Chewton Mendip Rising or simply Chew Head, this strong spring is the source of the River Chew and is reached by an elevated walkway alongside the nascent river. There are in fact two springs, within a few metres of each other, which emerge from the White Lias limestone and drain Chew Down to the west of Chilcompton and Ston Easton. On the hill to the south there is a 7m deep brick-lined pit which is said to act as an overflow during high flow. The river is a vital water supply, feeding two Bristol Waterworks reservoirs at Litton, and ultimately Chew Valley Lake, the South West's largest reservoir. Severe concerns have been raised about the potential impact of fracking in the spring's catchment area, which some believe threatens to have a serious detrimental effect on the entire region's water supply.

Ref: N. Barrington, W. Stanton, The Complete Caves and a view of the hills, p 51 (1977)

Chewhill Head Rising. Photo by Rob Taviner

FORD SPRINGS
Springs

Ford Springs is the collective name for a series of small springs clustered around the foot of Watery Combe which feed into the Bristol Waterworks 'Line of Works'. **Lower Street Spring** (ST 5943 5352) rises from Dolomitic Conglomerate in a large stone chamber, as does **Ford House Spring** (ST 5923 5355). **Watery Combe Spring** (ST 5918 5351 Alt 128m) wells up in the floor of the lowest of four shafts located in the valley itself, while the larger **Watery Combe Tunnel Spring** (ST 5916 5347 Alt 130m) rises in the floor of the brick-lined tunnel 60m down-valley of the second shaft in the valley. The tunnel continues right through to the first shaft where it is bricked up. The latter spring is said to have been dye-tested from water sinking in the vicinity of Grig's Pit Spring. There are many other small springs in the vicinity and a borehole was sunk at nearby Pullin's Dairy.

Ref: W. I. Stanton, Logbook 14 pp 43-44 (1983)

GRIG'S PIT SPRING
Spring
ST 5889 5315 Alt 140m

Located near the foot of the impressive rocky ravine which runs down Burges's Combe and through Grig's Pit Wood, this intermittent tufaceous spring feeds into the old Bristol Waterworks 'Line of Works' in Watery Combe. The presence of charcoal suggests that the valley was once used for smelting.

Ref: W. I. Stanton, Logbook 14 p 42 (1977)

WILLET'S LANE HOLE
Cave
ST 5887 5301 L 44m VR 20m Alt 158m

Located in Grig's Pit Quarry and also known as Ife Hole and Grig's Pit Wood Cave, this cave consists of two entrances (an upper and a lower) which unite in a roomy solutional passage. A twisting descent through shored boulders can be pursued to a small chamber which terminates in a tiny, intermittent sump. Christened Willet's Lane Hole by Balch and Howard Kenney in 1947, it was apparently exposed by quarrying in 1931 and first noted by cavers in 1938. The BEC dug into a lower level chamber and side passage in 1950 and renamed it Ife Hole, and it has been probed at various times since, most notably by MNRC who excavated a further rift and small chamber to reach the current lowest point in 2014-18. Like the sinkholes at Cutler's Green it may have begun life as a hydrothermal upwelling.

Ref: N. Barrington, W. Stanton, The Complete Caves and a view of the hills, p 175 (1977) & A. Watson, Personal communication

GRIG'S PIT QUARRY CAVE
Cave
ST 5885 5300 L 3m VR 1m Alt 158m

Located in the same small quarry as the larger Willet's Lane Hole, a low phreatic arch enters a small chamber containing a calcite vein.

HOLE IN THE FIELD
ST 584 529

Cave

L 15m VR 10m Alt 187m

No sign now remains of this large field collapse, which yielded a 10m pitch into a narrow, muddy rift. Explored by the WCC in 1983, it swiftly choked in both directions. It has since been filled in and the NGR is an approximation.

Ref: WCC Logbook 1981-83 pp 149-150 (1983)

CHEDDAR ROAD SHAFT
ST 5835 5255

Mine

L 1m VR 1m Alt 194m

Located in bushes by the side of the lane, this solitary ginged mineshaft, appears to be an isolated 17th century lead trial.

Ref: C. North, Personal communication

CHEWTON MENDIP AU BASE
ST 591 526 approx.

Mine

Marked on the map at the start of Somerset v Hitler, the location for this secret wartime AU base was pointed out by Waldegrave estate worker Charlie Ford who described '*an old mineshaft in the woods*'. With the aid of the Cranmore soldiers they dug a base that went straight down and then horizontal. It was floored with duck boards but was very damp. The site subsequently collapsed and can no longer be recognised. During WWII a steam waggon manufactured at Cutler's Green (see below), was allegedly used as a kit store by the Nedge Home Guard, who operated an observation hill on nearby Nedge Hill. They called themselves the Free French after their Commanding Officer, Lt Douglas, who had managed to escape from France by the skin of his teeth.

Ref: D. Brown, Somerset v Hitler, Countryside Books p 61 & p 77 (1999)

HELLARD'S CAVE
ST 5922 5238

Cave

L 24m VR 5m Alt 174m

This cave, formed entirely in Dolomitic Conglomerate, was opened by a mechanical digger in 1976, during excavation for a swimming pool. It consisted of an igloo chamber, described as '*big enough for ten people*', with a decorated bedding plane extending '*almost 100 feet*', to the south-east. Also known as both Plug Hole and Hellard Brothers Cave, it was permanently sealed within a few days.

Ref: N. Barrington, W. Stanton, The Complete Caves and a view of the hills, p 95 (1977) & Stanton Logbook 13 p 162

CUTLER'S GREEN SINKHOLES

Caves

This extraordinary site was uncovered following the sudden disappearance of a carp pond in 2003. During an attempt to discover the cause of the problem, the landowner stripped off the overlying soil to reveal an old wave-

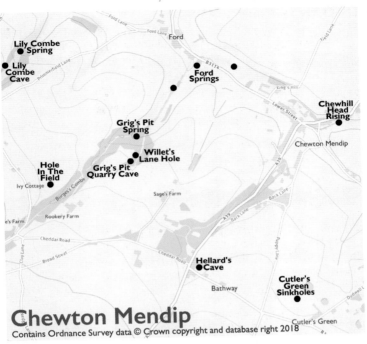

Chewton Mendip

Contains Ordnance Survey data © Crown copyright and database right 2018

cut platform, pierced by a collection of large solutional holes, filled with ochreous deposits. These probably began life as hydrothermal upwellings (possible hot springs) which were subsequently enlarged by surface drainage. Most of the holes have been at least partially excavated and the landowner has recovered an amazing collection of geological and palaeontological material from the surrounding land, some of which is now in Wells and Mendip Museum. Of additional interest, is the fact that Cutler's Green is the site of an old steam-powered iron foundry which produced swords for the British Army before expanding into the manufacture of motor vehicles. The Mendip Motor and Engineering Works produced steam lorries and petrol vans, and even designed and built their own light car. The Cutler's Green works is thought to have produced over 500 vehicles, of which only one incomplete example survives.

Ref: E. Sandford, Cutler's Green Sinkholes, BEC Belfry Bull 545 pp 14-18 (2013) & E. Sandford, Diggers to Excess...!, BEC Belfry Bull 556 pp 18-24 (2015)

CUTLER'S GREEN SINKHOLE 1
(ST 5977 5216) L 10m VR 10m Alt 173m - The first of the shafts to be properly examined, this is a fluted shaft that becomes too tight to follow. There are stemples in situ.

CUTLER'S GREEN SINKHOLE 2
(ST 5977 5216) L 8m VR 4m Alt 173m - This roomy cavity features an attractive rock bridge and two entrances. It soon becomes too tight to follow. There is a third entrance, which doesn't quite connect.

CUTLER'S GREEN SINKHOLE 3
(ST 5978 5216) L 6m VR 6m Alt 173m - This shaft terminates in an impassable bedding plane.

CUTLER'S GREEN SINKHOLE 4
(ST 5978 5215) L 4m VR 4m Alt 173m - This short shaft is thought to be the hole down which the bulk of the carp pond vanished.

CUTLER'S GREEN SINKHOLE 5
(ST 5979 5216) L 10m VR 10m Alt 173m - This shaft gradually becomes smaller, although it has not yet been dug to a complete conclusion.

CUTLER'S GREEN SINKHOLE 6
(ST 5978 5216) L 4m VR 4m Alt 173m - Another short shaft where work remains incomplete. It may connect with Cutler's Green Sinkhole 5.

CUTLER'S GREEN SINKHOLE 7
(ST 5977 5216) L 3m VR 3m Alt 173m - A short shaft which may connect with Cutler's Green Sinkhole 1.

CUTLER'S GREEN SINKHOLE 8
(ST 5977 5218) L 150m VR 20m Alt 173m - By far the largest of the sinkholes, this roomy shaft was filled in c.2006, following a collapse which threatened to swallow the hedge. It was re-excavated by the landowner in 2015, which allowed the BEC to explore roughly 60m of upper level passages. In 2016, the landowner deepened the shaft, which opened access to a lower set of passages, which are actively being pursued (2019).

DUDWELL FARM SHAFT   Cave
ST 6163 5228   L 11m VR 8m Alt 182m
Surrounded by a broken-down fence, this small isolated cave, which is formed in Jurassic limestone, was first recorded by the MNRC in 1964. A 6m deep shaft yielded a steeply-descending passage terminating in a sump. Despite attempts to level the site, the shaft remains obstinately open.

Ref: Shaft at Chewton Mendip, The Mendip Caver 1(5) (1964)

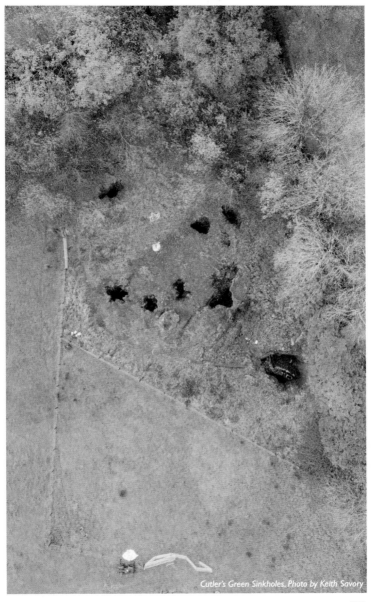

Cutler's Green Sinkholes. Photo by Keith Savory

ABBREVIATIONS

ACG	Axbridge Caving Group (& AS, with Archaeological Society)
AMCC	Aldermaston Mountaineering and Caving Club
AOD	Above Ordnance Datum (Above sea level)
AONB	Area of Outstanding Natural Beauty
ATLAS	Association of Thrupe Lane Advanced Speleologists
AWT	Avon Wildlife Trust
BCA	British Caving Association
BCRA	British Cave Research Association
BCSS	Beechen Cliff Speleological Society
BDCC	Bracknell District Caving Club
BEC	Bristol Exploration Club
BGS	British Geological Survey
BW	Bristol Water plc
CCG	Cotham Caving Group
CDG	Cave Diving Group
ChCC	Cheddar Caving Club
CPC	Craven Pothole Club
CSCC	Council of Southern Caving Clubs
CSS	Cerberus Speleological Society
EDCG	East Devon Caving Group
GSG	Grampian Speleological Group
HMSO	Her Majesty's Stationary Office
MCG	Mendip Caving Group
MCR	Mendip Cave Rescue (formerly **MRO** Mendip Rescue Organization)
MCRA	Mendip Cave Registry & Archive
MEG	Mendip Exploration Group
MHMCo	Mendip Hills Mining Company
MKHP	Mendip Karst Hydrology Research Project
MNRC	Mendip Nature Research Committee
NGR	National Grid Reference
NHASA	North Hill Association for Speleological Advancement
SAC	Special Area of Conservation
SANHS	Somerset Archaeological & Natural History Society
SBSS	South Bristol Speleological Society
SMCC	Shepton Mallet Caving Club
SMRG	Somerset Mines Research Group
SSSI	Site of Special Scientific Interest
SSSS	Sidcot School Speleological Society
SVCC	Severn Valley Caving Club
SWCC	South Wales Caving Club
SWT	Somerset Wildlife Trust
UBSS	University of Bristol Spelaeological Society
WCC	Wessex Cave Club
WNHAS	Wells Natural History and Archaeological Society
WWA	Wessex Water Authority

APPENDICES

APPENDIX I

MENDIP CAVE REGISTRY & ARCHIVE (MCRA)

The Mendip Cave Registry & Archive (MCRA) was formed in January 1956 with the object of recording and indexing all available information on caves, mines, rock shelters and other sites of speleological or underground interest in Somerset and the surrounding area. This information is freely available online at **www.mcra.org.uk**, and includes details on well over 2,500 sites. A registered charity, the membership of the group is drawn from volunteers representing nearly all the major clubs active in the region and is an independent and entirely self-financing body.

The website is constantly evolving, and as well as the main registry, includes a comprehensive bibliography and a growing list of online publications, including the logbooks of many famous individuals such as Tony Jarratt and Willie Stanton. Other topics covered include cave surveys (many of which are freely available for download), original and historic articles and films and audio recordings, and an ever-expanding database of photographs and postcards. These images are free for cavers to use in publications or on websites, providing the photographer and the MCRA are properly acknowledged. If a higher-quality image is required, please forward an email to the Archivist specifying the image reference number and a high-resolution image will be returned by email. An additional collection of images is available on the MCRA Flickr page (**www.flickr.com/mendipcaveregistryarchive**).

The MCRA also publishes books relating to caving, including *Mendip Underground* by Alan Gray, Rob Taviner and Richard Witcombe, *Earth Colours - Mendip and Bristol Ochre Mining* by Marie Clarke, Neville Gregory and Alan Gray, and Peter Burr's remarkable *Mines and Minerals of the Mendip Hills*. This has become a very successful method for publishing works which might otherwise be deemed un-commercial, and there are several more important volumes currently in the pipeline, including a further two instalments of *Somerset Underground*!

APPENDIX 2
CAVE CONSERVATION

From the very first moment someone enters a new cave passage, an irreversible process of change and deterioration begins. Some 'wear and tear' seems inevitable, but steps should always be taken to try and avoid unnecessary damage. A genuinely considerate attitude to the surroundings, coupled with a willingness to pass the message on to others, can go an awfully long way to minimising the impact. Marker tapes remain the commonest method of demarcating areas of vulnerable formations or floor deposits. If you are lucky enough to discover a new cave, try and establish an 'environmentally sensitive' path for others to follow. Initial taping should be done neatly and sympathetically, having regard to a sensible route for cavers to take in passing the area. Tapes will get muddy, broken or sometimes dragged out of place over time, and checks should be made to ensure that they are still fulfilling their intended role. Cave photographers, and their assistants, should take particular care when framing their photographs. Do NOT cross over or move conservation tapes, and always look behind you before moving backwards when seeking the right picture through the viewfinder or screen. Unfortunately, several notable formations have been damaged or destroyed in this manner. Cave diggers working underground can easily create a mess. Whilst this might be inevitable during the active phase of a dig, a good clean-up should take place after work has ceased. Plastic bags are often used for moving spoil, but ideally, they should be emptied in a chosen dump and eventually removed from the cave. It is all too common to see bags stacked in unsightly piles and left underground. In many cases dry stone retaining walls could have been quickly built instead and the spoil dumped behind them. If access and skills permit, it is good to use stone and concrete shoring, rather than steel scaffold poles, to stabilise loose areas. Stone and concrete have a very long life, whereas replacing rusty scaffolding in unstable chokes will be a hazardous chore for someone in the not too distant future. Surface dig sites are often highly sensitive and must never begin without first obtaining the permission of the landowner. Secure fencing is almost always needed, and all pits and shafts need safeguarding from inquisitive farm animals. Spoil dumps should be chosen carefully and again stone retaining walls are effective and look good. When work eventually ceases make sure the farmer or landowner is satisfied with your tidying up and entrance security. When digs 'don't go' it often happens that the team drifts away in piecemeal fashion leaving a very messy and sometimes unsafe site. There is no surer way to upset a farmer and give him a very good reason for saying 'No!' to the next explorer who seeks permission to dig on his land. The same concerns apply to other underground sites. Miners' stone walls, or 'deads', should not be disturbed or marked, and artefacts must not be tampered with or removed under any circumstances. Old wood in particular, should NOT be touched as it is liable to disintegrate, possibly with fatal results.

KEY POINTS

Adopt a positive attitude towards reducing deterioration of the cave environment.

Treat calcite formations with great care. Avoid breakage and muddying through unnecessary contact. DO NOT TOUCH!

Don't needlessly trample over undisturbed mud floors or sand and gravel banks. They may contain important sedimentary deposits.

Stick to the marker tapes. They are there to help preserve areas of particular interest or beauty.

Avoid leaving litter or food scraps underground or around the entrance. They look unsightly and are a source of pollution.

The use of acetylene lamps is no longer acceptable. Please use electric lighting.

Leave cave creatures undisturbed. All bats are protected species and it is unlawful to disturb or harm them in any way.

Features of archaeological interest should be left intact for detailed study by specialists.

Clubs should seriously consider taking a practical interest in conserving a specific cave by collecting litter, cleaning muddied formations, maintaining tape markers and removing graffiti.

Please remember that caving forms just one part of a wider countryside community. Dangerous car parking, leaving surface litter and damaging stone walls and fences are anti-social and damaging to the general environment. Such action will also upset landowners and farmers.

TAKE NOTHING BUT PHOTOGRAPHS

LEAVE NOTHING BUT FOOTPRINTS

and try and avoid leaving those if you can possibly help it!

APPENDIX 3
SAFETY

Unlike many sporting pastimes, caving offers considerable opportunity for doing irreparable damage to the participants. Analysis of the several hundred incidents the MCR has been involved in has identified the main causes of rescues to be the result of **FLOODS**, **FALLS** and **FAILURES** of crucial equipment. Obviously, no single set of rules can cover every eventuality but in general *Accidents Don't Just Happen*, and there is usually some error of judgement at the root cause of them. You can minimise the chances by applying the following basic procedures.

WARNING: FLOODING - numerous caves are liable to flooding, and up to date LOCAL weather information should be obtained prior to any trip. Don't set off down a wet cave when the forecast threatens heavy rain and try and seek advice from experienced cavers who know about local and prevailing stream flows.

Leave accurate and unambiguous written word of the locations you intend to visit, your intended route and a realistic indication of your expected time of return. Try and leave call-out messages with knowledgeable people, but if you do have to leave written messages with friends elsewhere, then please make them aware of the correct call-out procedure. Please do not change the basic plan and then neglect to amend the call-out messages left behind.

Never go underground alone.

Ensure that your equipment is serviceable. Wear clothing which is warm and suitable for the environment. Carry adequate food and fluids and try and eat beforehand.

Learn proper techniques and always use a lifeline on vertical or steep descents. Learn to tie knots correctly.

Regard all fixed aids with suspicion. Very few in situ bolts are maintained in any way and many have been in place for several years. They should be checked carefully. In many cases there are often perfectly adequate natural belays available nearby. **NEVER** stake your life on a single point of failure! Back up the main belay point to a safe and independent secondary.

Those contemplating using SRT are urged to seek expert training and advice.

Treat boulders with caution and stick to established and well-worn routes around or over them.

Don't loiter at the foot of a pitch. Stand clear of falling debris.

Understand the dangers posed by the presence of carbon dioxide. It is an insidious gas which can be dangerous (and indeed fatal) in high concentrations. It has no smell and can occur anywhere. Headaches and breathlessness are early warning signs. Be alert and if you suspect the presence of carbon dioxide, evacuate the site immediately. A CO_2 level of just 1% can lead to fatigue and clumsiness, while 2% can produce headaches, increased lung ventilation and a distinct loss of energy. It can take several days for the body to return to normal. Panting, dizziness and blurred vision occur at 3% and these symptoms increase dramatically from that point onwards. Prolonged exposure at 5% can lead to irreversible effects to health, while anything above 5% can lead to unconsciousness and death.

Learn basic rescue techniques, as it may very well be you who has to use them!

Learn *FIRST AID*. Injuries, even relatively minor ones become a much more serious matter when they occur underground. The right equipment may take hours to arrive, the hostile environment may compound the severity of the situation and evacuating an immobilised casualty can prove a serious undertaking. Much may depend upon the knowledge and ingenuity of those first on the spot. The carrying of a basic First Aid kit is strongly recommended. As a bare minimum, try to find room for a survival bag and a polythene-wrapped, prepared wound dressing.

APPENDIX 4
MENDIP CAVE RESCUE (MCR)

The MCR (formerly MRO), is the second oldest cave rescue organisation in the world and acting on behalf of the Police, is primarily responsible for rescues from caves and disused stone and mineral mines in the Bristol area, Somerset, Wiltshire and Dorset. The organisation maintains a large quantity of specialist rescue equipment, and is run by wardens, most of whom are locally resident cavers. It is their job to ensure that suitable call-out arrangements, equipment and manpower are available when needed. However, the organisation actively promotes self-help, an approach which involves all cavers and encourages them to assist fellow cavers in difficulty. The work of the MCR is voluntary and funded entirely by donations and support received from the whole of the caving community.

HOW TO CALL OUT A MENDIP CAVE RESCUE TEAM IN AN EMERGENCY

1. **Dial 999** (or 112 - the single European Emergency Number).
2. Ask for the **POLICE**.
3. If calling from a mobile ask for **AVON AND SOMERSET POLICE**.
4. Then ask for **CAVE RESCUE**.
5. Prepare to give the following details: Cave Name; Location within cave; Numbers and condition of those involved; Details of any injuries sustained.
6. **STAY BY THE PHONE**. If using a mobile stay in a good reception area.
7. A Mendip Cave Rescue Warden will call you back for more details so please keep the phone line free.

APPENDIX 5
CAVING ORGANISATIONS

Those intending to venture underground are strongly advised to contact specialist organisations such as caving clubs or mining societies. Many offer excellent accommodation and library facilities and can provide an invaluable point of contact with experts on the more technical aspects of underground exploration, such as cave diving, cave digging, surveying, archaeology, hydrology and biology. Membership fees are generally very modest and can include access to the British Caving Association's public liability insurance scheme which provides landowners with cover against claims by cavers up to a maximum of £5 million. Not surprisingly in this age of litigation, most owners will insist that visitors have an appropriate level of cover. Doing so without appropriate cover could seriously jeopardise relations with landowners' and in the event of a serious accident could put the whole future of caving at risk. Suitable local contact details can be found on the CSCC website **www.cscc.org.uk**. In addition to the specialist local organisations, there are national bodies that represent some of the broader issues.

BRITISH CAVING ASSOCIATION (BCA)
www.british-caving.org.uk
The BCA is the governing body for underground exploration in the UK. It represents all those persons and groups with a genuine sporting or scientific interest in caves and caving, and in close liaison with NAMHO (see below), also seeks to represent those with an interest in mines and the man-made underground environment. In a wider sense the organisation seeks to promote a better public understanding on all matters relating to underground exploration.

BRITISH CAVE RESEARCH ASSOCIATION (BCRA)
www.bcra.org.uk
The BCRA exists to promote the study of caves and associated phenomena. It attains this object by supporting cave and karst research, by encouraging original exploration (both in the UK and on overseas expeditions) and by collecting and publishing speleological information. It supports a number of Special Interest Groups that allow BCRA members to associate and communicate in a more detailed way across a range of specialist subjects, including Archaeology, Cave Biology, Cave Radio and Electronics, the Use of Explosives and Cave Surveying. Any caver with an interest in any of these topics is urged to contact the BCRA.

CAVE DIVING GROUP www.cavedivinggroup.org.uk
This is the national organisation for cave diving in Great Britain. It also undertakes cave diving training, and nobody should contemplate pursuing this particular specialist branch of caving without first contacting this organisation.

NAMHO www.namho.org
The National Association of Mining History Organisations is the national body representing all those interested in mining history in the UK and Ireland.

APPENDIX 6
FURTHER READING

In addition to the books and publications listed below, readers are urged to browse the various caving club journals, many of which are freely available online. Links to these can be found in the 'Links' section of the MCRA website (www.mcra.org.uk). For an overall picture, readers are strongly urged to refer to the comprehensive list of references available in the online bibliography.

M. Atkinson, **Exmoor's Industrial Archaeology**, Exmoor Books (1997)

T. Atkinson et al., **Mendip Karst Hydrology Research Project, Phases 1 and 2**, WCC Occasional Publications, Series 2 No. 1, WCC (1967)

R. Atthill, **Old Mendip**, David & Charles, Newton Abbot, Devon 2nd edition (1971)

R. Atthill, **Mendip - A New Study**, David & Charles, Newton Abbot, Devon (1976)

E. A. Baker and H. E. Balch, **Netherworld of Mendip**, J. Baker and Son (Clifton), Simpkin, Marshall, Hamilton, Kent and Co. (London) (1907)

H. E. Balch, Mendip — **The Great Cave of Wookey Hole**, John Wright and Sons Ltd (Bristol) Simpkin, Marshall (1941) Ltd. London, 3rd edition (1947)

H. E. Balch, Mendip — **Cheddar, Its Gorge and Caves**, John Wright and Sons Ltd (Bristol), Simpkin, Marshall (1941) Ltd. London, 2nd edition (1947)

H. E. Balch, **Mendip — Its Swallet Caves and Rock Shelters**, John Wright and Sons Ltd (Bristol), Simpkin, Marshall (1941) Ltd. London, 2nd edition (1948)

N. Barrington and W. I. Stanton, **Mendip — The Complete Caves and a View of the Hills**, Barton Publications, Cheddar Valley Press, Cheddar, Somerset, 3rd revised edition (1977)

K. Branigan and M. J. Dearne, **A Gazetteer of Romano-British Cave Sites and their Finds**, Dept. of Archaeology and Prehistory, University of Sheffield (1991)

D. Brown. **Somerset v Hitler, Secret Operations in the Mendips 1939-1945**, Countryside Books (1999)

P. Burr, **Mines and Minerals of the Mendip Hills**, Volumes I & II, MCRA (2015)

M. Clarke, N. Gregory and A. Gray, **Earth Colours - Mendip and Bristol Ochre Mining**, MCRA (2012)

M. Crocker, **Cheddar Gorge Climbs** (2015)

M. Crocker, **Avon Gorge, A Climbers' Club Guide**, The Climbers' Club (2017)

W. B. Dawkins, **Cave Hunting**, Macmillan and Co. London (1874) Republished by E. P. Publishing Ltd, Wakefield, Yorkshire (1973)

D. P. Drew et al., **Mendip Karst Hydrology Research Project, Phase 3**, WCC Occasional Publications, Series 2 No. 2, WCC (1968)

K. L. Duff et al., **New Sites for Old, A Student's Guide to the Geology of the East Mendips**, Nature Conservancy Council (1985)

A. Farrant, **Eastern Mendip**, British Geological Survey (2008)

A. Farrant, **Western Mendip**, British Geological Survey (2008)

J. W. Gough, **The Mines of Mendip**, OUP (1930) Revised edition David & Charles, Newton Abbot (1967)

A. Gray, R.M. Taviner and R.G. Witcombe, **Mendip Underground**, MCRA (2013)

J. Hamilton and J. F. Lawrence, **Men and Mining on the Quantocks 2nd Edition**, Exmoor Mines Research Group (2008)

J. D. Hanwell and M. D. Newson, **The Great Storms and Floods of July 1968 on Mendip**, WCC Occasional Publications, Series 1 No. 2, WCC (1970)

A. L. Holt, **West Somerset, East Somerset & Old North Somerset, Romantic Routes & Mysterious Byways**, Charles Skilton Ltd (1984, 1986 & 1987)

Dom. E. Horne, **Somerset Holy Wells and Other Named Wells**, The Somerset Folk Series No. 12 (1923)

P. Johnson, **The History of Mendip Caving**, David & Charles, Newton Abbot, Devon (1967)

M. H. Jones, **The Brendon Hills Iron Mines and The West Somerset Mineral Railway**, A New Account, Lightmoor Press (2011)

F. A. Knight, **The Sea-Board of Mendip**, J. M. Dent & Company (1902)

F. A. Knight, **The Heart of Mendip**, J. M. Dent & Company (1915)

R. Legg, **Steep Holm Legends and History**, Wincanton Press (1993)

N. Macmillen & M. Chapman, **A History of the Fuller's Earth Mining Industry Around Bath**, Lightmoor Press (2009)

M. C. McDonald and D. M. Price, **Somerset Sump Index**, CDG (Somerset Section) (2008)

D. P. Mockford and A. J. Male, **Caves of the Bristol Region**, The Bristol Cave Register (1974)

P. Quinn, **Holy Wells of Bath & Bristol Region**, Logaston Press (1999)

J. Rattue, **The Living Stream, Holy Wells in Historical Context**, Boydell Press (1995)

L. Richardson, **Wells and Springs of Somerset**, HMSO (1928)

J. Rutter, **Delineations of the North Western Division of the County of Somerset**, Longman Rees & Co. (1829)

J. Savory, **A Man Deep In Mendip, The Caving Diaries of Harry Savory 1910-1921**, Alan Sutton, Gloucester (1989)

V. Simmonds, **An overview of the archaeology of Mendip caves and karst**, www.mcra.org.uk (2014)

D. I. Smith (Ed), **Limestones and Caves of the Mendip Hills**, David & Charles, Newton Abbot, Devon (1975)

R. Thomas, Quarry Faces, **The Story of Mendip Stone**, Ash Tree Publications (2015)

S. Watson, **Secret Underground Bristol**, The Bristol Junior Chamber (1991)

R. G. Witcombe, **Who Was Aveline Anyway? - Mendip's Cave Names Explained**, WCC Occasional Publication (2008)

ACKNOWLEDGEMENTS

The author would like to thank the many dozens of Mendip cavers whose contributions have made this volume possible. This continuing collective effort only serves to underline the special and enduring character of the Mendip caving community.

The daunting task of proof reading the book has been carried out with great care and diligence by Mandy Voysey, Nick Richards and Vince Simmonds, all highly-regarded local cavers who have a great deal of technical and local knowledge. It is a vital task in the production of any guide and my grateful thanks go to all of them.

Special thanks must also go to the following people and organisations (listed in alphabetical order) whose contributions have proved invaluable.

Shaun Anning, James Begley, John Bennett, Tony Boycott, Ambrose Buchanan, Chris Castle, Tom Chapman, Nick Chipchase, Hayley Clark, Kevin Clinton, Ray Deasy, Cliff Dockrell, Pete Flanagan, Stu Gardiner, Alex Gee, Emma Gisborne, Peter Glanvill, Martin Grass, Alan Gray, Robin Gray, Nick Harding, Doug Harris, Tom Harrison, Nick Hawkes, Andy Hebden, Phil Hendy, Trevor Hughes, Graham 'Jake' Johnson, Dave King, Kate Lawrence, Daniel Medley, Martin Mills, Alison Moody, Andy Morgan, Mike Moxon, Graham Mullan, Richard 'Spike' Neal, Clive North, Mick Norton, Dennis Parsons, Chris Pollard, Brian Prewer, Duncan Price, Chris Richards, Nick Richards, Pete Rose, Estelle Sandford, Keith Savory, Danielle Schreve, Sue Shaw, Vince Simmonds, South West Heritage Trust, Adrian Vanderplank, Mandy Voysey, Matt Voysey, Dave Walker, Ed Waters, Andy Watson, John 'Tangent' Williams, Rich Witcombe, Wookey Hole Caves

INDEX

1921 DIG 182
200 YARD DIG 176

ACG-01 to 27 (Sandford Hill) 84-89
ADAM'S WELL 77
ALCO HOLE 81
ALL EIGHTS FIELD SHAFTS 224
ALL EIGHTS MINESHAFT 223
AQUEDUCT HOLE 207
ARCHWAY CAVE 54
ARM COVER LEAT 222
Arsey Pit I 120
Arsey Pit II 121
ASH TREE PIT (Shipham) 154
ASH TREE RIFTS (Burringon) 168
AUGER'S MINES 89
AVELINE'S HOLE 174
Axbridge Bypass Cave 139
AXBRIDGE CHURCH WELLS 151
AXBRIDGE HILL CAVERN 144
AXBRIDGE HILL OCHRE PITS 140
Axbridge New Cave 137
AXBRIDGE OCHRE CAVERN 147
AXBRIDGE OCHRE LOWER MINE 145
AXBRIDGE OCHRE MIDDLE MINE 145
AXBRIDGE RISING 151
AXBRIDGE STATION WELL 150

BALCONY CAVE 147
BANWELL BONE CAVE 70
BANWELL COLLAPSE 74
BANWELL HIGH STREET MINE 75
BANWELL LEVVY 73
BANWELL OCHRE CAVES 79-81
BANWELL OCHRE PIT 79
BANWELL RISING 76
BANWELL STALACTITE CAVE 71
BARLEYCOMBE WOOD SPRING 59
BARREN HOLE 177
BARROW WELL 205
BARTON DROVE SHAFT 133
BARTON SHELTER 133
BAT HOLE 86
BATCH WELL 196
BATH SWALLET 169
Bathing Pool Swallet 169
BATTERY CAVE 35
Bear Pit 145
BEARD'S SECOND CAVERN OF BONES (1833) 51
Beaumont's Hole 211
BEDROCK PIT 56
Bedstead Rift 121
BEECHES HOLE 132
BELT SINK NORTH 219
BELT SINK SOUTH 219
Big Cave 79
BIGHORN RIFT 107
BISHOP'S WELL 110
BLAGDON BONE FISSURE 200
BLEADON BONE CAVE 61
BLEADON CAVERN 52
BLEADON QUARRY CAVES 60-62
BLEADON QUARRY WELL 61
BLEADON SPRINGS 62
BLIND PIT (Hutton) 56
BLIND PIT (Shipham) 120
BOOT MINE 119
BORDERLINES DIG 210
BOS SWALLET 169
BOULDER SHAFT (Burrington) 186
BOULDER SHAFT (Shipham) 121
BRAMBLE SHAFT 121
BREAN DOWN SEA CAVES 28-36
BRIDEWELL (Churchill) 110
BRIDEWELL (Elborough) 58

BRIDEWELL CAVE (Banwell) 70
BROWN'S WELL 123
BRUCE'S HOLE 177
BUDDLE'S WOOD OUTER SHAFT 232
BUDDLE'S WOOD SHAFTS 1-9 229-232
BURRINGTON HAM MINE 186
Burrington Hole 180
BURRINGTON RAKES 195
Burrington Waterworks Adit 191
BUTTER WELL 123
BYPASS GEODE 139

CAFE HOLE 171
CALCITE SHAFT 110
CALLOW CAVE 148
CALLOW HILL QUARRY BOREHOLE 152
Callow Limeworks Cave 153
CALLOW LIMEWORKS QUARRY CAVE 154
CALLOW QUARRY BEDDING PLANE DIG 153
CALLOW ROCK QUARRY CAVE 154
CALLOW ROCKS CAVE 153
CALLOW SLIT 148
CANADA COMBE CAVE 43
CANADA COMBE FISSURE 43
CANADA COMBE ROCK SHELTER 46
CANYON SPRING 221
Capel's Hill Ochre Mine 148
Capon Mine 222
CAR PARK CAVE 153
CARCASS CAVE 136
CEMETERY SHAFT 122
CHAPEL RIFT 89
CHARD'S WELL 148
CHEDDAR ROAD SHAFT 236
CHEDDAR WOOD OCHRE PITS 150
Cherry Tree Mine 89

Chew Head 234
CHEWHILL HEAD RISING 234
CHEWTON MENDIP AU BASE 236
Chewton Mendip Rising 234
CHIMNEY POT 220
CHIMNEY SINK 220
CHOCOLATE WALL SINK 220
CHRISTON SPRING 59
Churchill Bone Cave 105
CHURCHILL CAVE 158
Churchill Swallet 108
CLASSIC SHAFT 102
CLAY HOLES 233
CLIFF QUARRY HOLE 204
CLIFF QUARRY RIFT 204
COFFEE POT 98
COLDWELL 68
COLE CHUTE 54
COLLARED PIT 121
COMBE LODGE RISINGS 197
COMPTON BISHOP OCHRE ARCH 128
COMPTON BISHOP OCHRE ARCH 2 128
COMPTON BISHOP SPRING 127
COMPTON COMBE MINESHAFT 204
COMPTON COMBE SHELTER 204
COMPTON MARTIN OCHRE MINE 204
COMPTON MARTIN RIFT CAVE 203
CORAL CAVE 134
COX'S WELL 111
CRAG HOLE 67
CROAKERS DROP 58
Croc Pit 56
CROCUS HOLE 119
CROOK PEAK HOLE 133
CROOK PEAK RIFT 133
CROSS PLAIN FISSURE 134
CROSS QUARRY CAVE 135
CROSS QUARRY GEODE 135

Index

CROSS SPRING 135
CROW CATCH 154
CROWN INN OCHRE MINE 151
CROWN INN SWALLET 108
CUTLER'S GREEN SINKHOLES 236-238
CUTTING CAVES & RIFT 153
Cyclops Cave 28

DAFFODIL MINE 118
Daffodil Valley Cave 116
DEEP GROOVE MINE 101
DENNY'S HOLE 128
Devil's Cave 127
DGW MINE 116
DICK TURPIN'S CAVE 139
DIPPING WELL 205
DITCH CAVE 30
DOLEBURY CAVERN 159
DOLEBURY HILL CAVE 160
DOLEBURY LEVVY 159
DOLEBURY WARREN MINESHAFT 160
DOLEBURY WARREN SWALLET 160
DOUBLE SPRING 221
DRAUGHTING PIT 56
Dreadnaught Holes 194
DRUNKARD'S HOLE 169
DUDWELL FARM SHAFT 238
DUNNETT SPRINGS 126
DUTTON'S SPRING 34

E HOLE 186
EAKER HILL FARM MINESHAFT 225
EAST TWIN MIDDLE SINK 194
EAST TWIN SWALLET 194
EAST TWIN TOP SINK 194
EAST WELL 112
EASTWOOD MANOR MINES 222
Egg Hole 96
ELBOROUGH MINE 58

ELEPHANT CAVE (Sandford Hill) 104
ELEPHANT'S HOLE (Burrington) 186
ELEPHANT'S RIFT (Burrington) 186
ELLENGE RISING 149
ELLICK FARM SWALLET 187
ELLICK SPRINGS 187
ELLIS' WONDER 90
Engine Pit 224

F & B HOLE 225
FAIRY TOOT 165
FENCE PIT 56
FENNEL HOLE 139
FERN MINE 95
Fester Hole 177
FIDDLER'S BAY CAVE 34
FIDLING'S WELL 111
FIG TREE MINE 150
FIVE SPRINGS 111
FLANGE SWALLET 190
FLAT BOTTOMED SINK 220
FLINT CREVICE 145
Footprint Cave 77
FORD HOUSE SPRING 235
FORD SPRINGS 235
FORTY ACRES MINE 148
FOX HOLES (Read's Cavern) 170
FOX'S HOLE (Compton Bishop) 129
FOXES HOLE (Burrington) 182
FRANCES PLANTATION SWALLET 219
FROG'S HOLE 183
FRY'S HILL PIT 150
FRY'S HILL SHAFTS 147-148
FULLER'S POND 112

GALLERY PIT CAVE 54
Gaping Ghyll 122
GARROW BOTTOM SPRING 221
GARROWPIPE FISSURES 214
GARROWPIPE SPRING 1 214
GARROWPIPE SPRING 2 215
GAUGE HOUSE RISING 196
GIANT'S CAVE 30
GIBBETS BROW SHAFT 210
GIRL'S HOLE 99
GLEBE WOOD SPRING 57
GOATCHURCH CAVERN 193
Goat's Hole 193
Goechurch Cave 193
GOODING'S 1 213
GOON'S HOLE 180
GRAN'S DISCOVERY 90
GREAT MAPLE MINE 73
GREAT RIFT 30
Great Spring (Axbridge) 151
Great Spring (Langford) 163
GREENDOWN MINESHAFT 233
GRIG'S PIT QUARRY CAVE 235
GRIG'S PIT SPRING 235
Grig's Pit Wood Cave 235
GROOF HOUSE SHAFTS 94
GUY HOLE 193

HALE WELL 113
HALF TIDE ROCK CAVE 36
HAM MINE 1 to 3 115
HARPTREE COMBE FIELD COLLAPSE 209
HARPTREE COMBE MINES 1 to 11 207-209
Harptree Combe Resurgence 207
HARPTREE COMBE SEEPAGE 209
HARPTREE COMBE WATERWORKS TUNNEL 207
HARPTREE COURT TUNNEL & GROTTO 221
HARPTREE FARM MINE 221

Harptree Wishing Well 207
HARPTREE WOODS SINK 219
HAUL ROAD SHAFT 154
HAWKESWELL QUOIT 187
HAWTHORN HOLE 218
HAY WOOD RIFT 42
HAY WOOD ROCK SHELTER 42
HAY WOOD SPRING 42
HAYES RIFT MINE 124
HAZEL MANOR MINE 203
HAZEL MINE (Shipham) 119
HEAL'S OCHRE MINES 108
HELICTITE WELL 125
HELLARD'S CAVE 236
HELLENGE HILL HOLE 63
HIGH MINE 106
HIGHER LEAZE MINES 59
HILLEND BARYTES MINE 70
HILLEND HOLE 69
HILLGATE HOUSE HOLE 63
HILLSIDE COTTAGE SEEPAGE 42
HILLSIDE TUNNEL 150
HINTON'S HOLE 70
HOLE IN THE FIELD 236
HOLLYBUSH SHAFT (Banwell) 72
HOLLY BUSH SHAFT (Shipham) 126
HOLLY FISSURE (Shipham) 119
HOLLY SHAFTS (Sandford Hill) 87
HOLLY WELL 155
HONORARY HOLES 153
HUTTON CAVERN (1757) 47
HUTTON CAVERN (1828) 49
HUTTON CAVERN 2 (1757) 49
HUTTON CAVERN DIG 1 56
HUTTON CAVERN DIG 2 57
HUTTON HILL HOLE 42
HUTTON OCHRE CAVE 57
HUTTON SHAFT 1 56
HUTTON SHAFT 2 56
HYWEL'S HOLE 214

Ife Hole 235
Impenetrable Fissure 214
IVY FISSURE 119

JAY'S CAVE 63
JOHNNY NASH'S HOLE 188

Keltic Cavern 170
Kentish Cave 144
KING MINE 83
KING'S CASTLE WELL 207
KNOWLE CAVE 110
KNOWLE WOOD MINE 110

LADIES WELL 57
LAMB BOTTOM ADIT 214
LAMB BOTTOM FISSURE 214
LAMB BOTTOM RIFT 213
Lamb Cave 211
Lamb Lair 211
LAMB LEER CAVERN 211
LAMB LEER QUARRY RIFT 210
LANGFORD RISING 163
LARGE CHAMBER CAVE 140
LAY-BY DIG 180
LETTERBOX CAVE 145
LILY COMBE CAVE 233
LILY COMBE SPRING 233
LIONEL'S HOLE 180
Little Wood Mine 101
LIZARD HOLE 187
LOOKOUT CAVE 30
Lost Cave of Axbridge 140
Lost Cave of Blagdon 200
Lost Cave of Burrington 1 165
Lost Cave of Burrington 2 174
Lost Cave of Burrington 3 182
Lost Cave of Burrington 4 168
Lost Cave of Burrington 5 183
Lost Cave of Churchill 109
Lost Cave of Harptree 213

Lost Cave of Hutton 46
Lost Cave of Loxton 64
Lower Garrow Cave 214
LOWER HARPTREE COMBE FISSURES 207
LOWER STREET SPRING 235
LOWER WELL 200
LOXTON CAVE 67
LOXTON CAVERN 64
LOXTON QUARRY CAVES 64
LOXTON SAND QUARRY CAVES 67-68
LOXTON SETT 68
LUDWELL CAVE 43
LUDWELL RISING 43
LYNCOMBE MINE 99
LYNCOMBE SHAFT 99

Main Entrance 81
MANGLE HOLE 87
MAX MILL 111
MAY TREE CAVE 52
MCRA-BD-1 to 56 28-36
MCRA-SH-1 to 14 (Sandford Hill) 83-98
MERECOMBE WOOD SHAFT 202
Middle Series 81
MILLIAR'S QUARRY CAVE 171
Mine 4001-4006 207-209
MINESHAFT 4092 225
MINESHAFTS 4113 to 4123 229-232
MNRC MINES (Sandford Hill) 90-96
MOAT CAVE 32
MOLE HOLE 187
MONARCH'S WAY SINK 221
MYATT'S MINE 139

NAMELESS CAVE 174
NEW HALL MINE 116
NEW SHAFT 119
NISSEN HUT CAVE 32
NORTH FIELD CAVE 154

OLD FORGE WELL 135
OLD SHAFT 2 122
ORCHARD HOLES 195

P HOLE 168
PALLET SHAFT 121
PARALLEL GASHES 160
Pearce's Pot 169
PEARL MINE 96
Pendarves Shaft 124
Peter Bird's Dig 190
Phelps' Hole 128
PICKEN'S HOLE 132
PIERRE'S POT 188
PIPSQUEAK HOLE 187
PITT FARM SPRING 218
PLANTATION HOUSE DRIP WELL 74
Plantation Mines 147-148
Plug Hole 236
Plumley's Den 182
PLUMLEY'S HOLE 171
POINT CAVE 32
POOL CAVES (Brean Down) 30
Pool Mine (Banwell) 79
POOL SINK (Harptree) 219
PORTCH'S POT 195
PRIMROSE CAVE 52
PSEUDO NASH'S HOLE 188
PSEUDO UDDERS HOLE 83
PURN HILL CAVE 60
PURN HILL MINES 60
PYLE WELL 115

QUARRY CAVE 176
QUIDDLER'S ARCH 190

RABBIT HOLE 96
RADFORD FARM SINK 233
RAG SPRING 205
READ'S CAVERN 170
REINDEER RIFT 28
RICHMONT CASTLE RIFT 209
RICKFORD FARM CAVE 195
RICKFORD FOX HOLE 195
RICKFORD RISING 196
RIDGE LANE SWALLET 210
RIFT MINE 208
ROAD ARCH 183
ROBERT'S PARADISE 102
ROCK GIRT PIT 98
ROD'S POT 169
ROOKERY FARM SPRING 202
ROSE WOOD BOUNDARY CAVE 137
ROWBERROW BEACH SHAFT 122
ROWBERROW BOTTOM SHELTER 161
ROWBERROW CAVERN 161
ROWBERROW FOREST IRON PITS 163
ROWBERROW FOREST MINE 162
ROWBERROW FOREST SWALLET 162
ROWBERROW LANE MINE 123
Rowberrow Shaft 120
ROWBERROW SWALLET 161
ROWBERROW WATERWORKS TUNNEL 158
RUBBISHY RIFT 116
RUSHY GROUND SWALLET 233

Index

SAND CAVE 77
SANDFORD BONE FISSURE 105
SANDFORD EAST OCHRE MINE 99
SANDFORD GULF 94
Sandford Hill Mine 1 96
SANDFORD HILL MINE 2 100
SANDFORD HILL MINE 3 100
SANDFORD HILL MINE 4 105
SANDFORD LEVVY 84
SANDFORD OCHRE CAVE 107
SANDFORD QUARRY BADGER CAVES 107
SANDFORD QUARRY CAVE 106
SANDFORD QUARRY CAVE 2 107
SANDFORD QUARRY MINE 98
SANDFORD RAKE 102
Sandford Rifts West 102
SANDFORD RIFTS EAST 100
SANDY CAVE 127
SAVILLE ROW SHAFTS 90-93
SAYE'S LANE RISING 163
SCHOOL HOUSE WELL 57
SCHRAPNEL HOLE 56
SCRAGG'S HOLE 129
SCUM HOLE 95
SECRET TUNNEL 32
SHAFT 120 93
SHEPHERD'S SHAFT 126
Shepton Secret Dig 210
SHERBORNE SPRING 222
SHERBORNE SPRING COLLAPSE 222
SHIPHAM GORGE QUARRY CAVE 155
SHIPHAM GORGE QUARRY FISSURE 155
SHIPHAM HILL QUARRY ALCOVE 155
SHIPHAM HILL QUARRY RIFT 155
SHIPHAM MINE 124
Shipham Old Mine Well 125
SHIPLATE FISSURE 64

SHIPLATE SPRING 63
Shirborne Spring 196
SHUTE SHELVE CAVERN 137
SHUTE SHELVE QUARRY HOLE 139
SIDCOT SCHOOL SHAFT 112
SIDCOT SCHOOL WELL 112
SIDCOT SWALLET 191
SILAGE TRACTOR COLLAPSE 215
SINGING RIVER MINE 125
SKELETON RIFT 101
SLAG HEAP SINK 220
SMALL PIT 56
SMCC-1 to 20 (Sandford Hill) 83-96
SMITHAM HILL COLLAPSE 215
SMITHAM HILL RIFT CAVE 215
SMITHAM SINK 220
Snogging Hole 180
SOFT WELL 229
SOUTH CAVITY 67
SOUTH QUARRY CAVE 41
SOUTH RIFT 101
Southall's Hole 129
SPAR POT (Burrington) 194
SPAR SHAFT (Sandford Hill) 106
Spider Holes 170
SQUIRE'S WELL 197
SSSS MINES (Sandford Hill) 89-107
ST MICHAEL'S TUNNEL 149
STANTON SHAFT (Shipham) 120
STANTON'S SHAFT (Sandford Hill) 100
STAR MINE 1 120
STAR MINE 2 121
Star Ochre Mine 99
STAR SHAFT 120
STOBART'S HOLE 155
STREET END HOLE 200
SUMMIT PLANTATION MINE 102
SUNNY COVE HOLE 62
SUPRA-AVELINE'S SHAFT 176
SUPRA-SANDY HOLE 128

SUPRA-SCRAGG'S HOLE 129
SWAN INN SWALLET 161
SWANCOMBE HOLLOW DIG 201
SWANCOMBE WOOD HOLE 201
**SWARBRICK MINES
(Sandford Hill)** 83-106

TEST MINE 106
THE BACKDOOR 144
The Levvy 84
THE MYSTERIOUS UPHILL BULGE 41
The Redding Works 204
The Rift 139
THE TWINS 145
Three Entrance Mine 107
TIMS WELL 200
TIN CAN ALLEY (Banwell) 74
TIN CAN ALLEY (Shipham) 116
TINKER'S CAVE 58
TOAD PIT (Axbridge) 144
TOAD'S HOLE (Burrington) 183
TOWER MINE 73
TOWER MINE 2 73
TOWERHEAD HOLE 81
TRAMWAY PITS 144
TRAT'S CRACK 182
TREE STUMP SHAFT 56
TRIPLE HOLE (Sandford Hill) 102
TRIPLE-H CAVE (Axbridge) 144
T-SHAPED GRUFF 119
TUMBLEDOWN WALL HOLE 57
TUNNEL BEDDING PLANE 176
TUNNEL CAVE 176
TWEEN TWINS HOLE 177
Twin Passage Mine 208
TWIN TUBES 62
TWO TREES PIT 56
Twyford Dig 57
TYRE PIT 56

UBLEY CONDUIT 202
UBLEY HILL FARM RIFT 203
UDDER'S HOLE 100
UNDER SHED PIT 56
UPHILL CAVERN 37
UPHILL QUARRY CAVES 37-40
UPPER BARTON SHELTER 133
UPPER CANADA CAVE 54
Upper Garrow Cave 214
UPPER ROSE WOOD BOUNDARY CAVE 137

VIXEN'S HOLE 58

WALDEGRAVE ARMS MINE 222
WALLABY HOLE 120
WATER VALLEY SWALLET 163
WATERY COMBE SPRING 235
WATERY COMBE TUNNEL SPRING 235
WEBBINGTON QUARRY CAVE 132
WELL SHAFT CAVE 55
WEST END OLD SHAFT 1 224
WEST END OLD SHAFTS 2 225
WEST END SINK 229
WEST END SPRING 229
WEST TWIN BROOK ADIT 191
WEST TWIN BROOK SPRING 191
WHISPERING GALLERY 90
WHITCOMBE'S HOLE 177
WHITE CLIFF CAVE 134
WHITLEY HEAD RIFT 70
WILLET'S LANE HOLE 235
WINKS WELL 147
WINSCOMBE RAILWAY CUTTING CAVES 113
WINTERHEAD HILL SWALLET 123
WINTERHEAD MINE 124
WINTERHEAD SHAFT 124
WINTERHEAD SWALLET 123
WINTHILL SHAFT 77

WOLF DEN 133
Womble Well 125
WOODBOROUGH GREEN 112
Wringstone Rocks Cave 133
WRINGSTONE ROCKS LEAD MINE 133
WURT PIT 218
Wyatt's Cave 195

YARBOROUGH FISSURE 78
YEW TREE SWALLET 191
YOGHURT POT 210

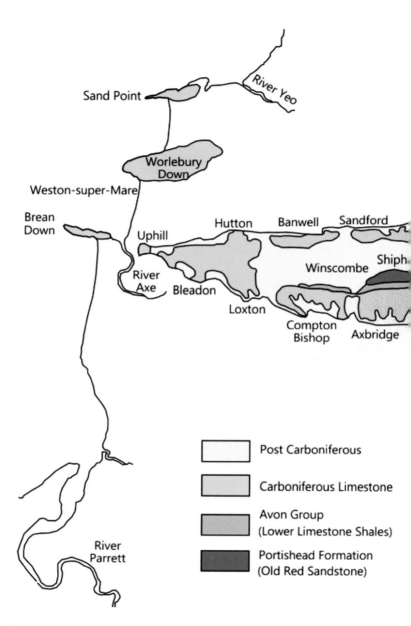

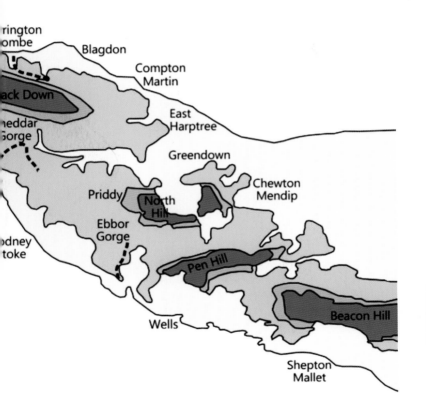

NOTES